THE AUTOBIOGRAPHY OF
Benjamin Franklin

Edited, with Introduction and Notes,
by Gordon S. Haight, Ph.D.

BENJAMIN FRANKLIN

Engraving by H. S. Hall

The Bettman Archive

The Autobiography
of Benjamin Franklin

Published for the Classics Club ® by

WALTER J. BLACK, INC. · ROSLYN, N. Y.

COPYRIGHT © 1941
BY WALTER J. BLACK, INC.
COPYRIGHT © RENEWED 1969

COPYRIGHTED UNDER THE UNIVERSAL COPYRIGHT CONVENTION
ALL RIGHTS RESERVED
PRINTED IN THE UNITED STATES OF AMERICA
TRADEMARK REGISTERED

List of Illustrations

Benjamin Franklin	*Frontispiece*
Title Page of Franklin's First Book: *A Dissertation on Liberty and Necessity, Pleasure and Pain*, 1725	65
A Threepence Note Printed by Franklin and Hall, 1764	85
Title Page of *Poor Richard's Almanac* for 1734	149
Title Page of *Plain Truth*, 1747	172
First Pages of *An Account of the New-Invented Pennsylvania Fire-Places*, Written and Printed by Franklin in 1744	183
First Page of *A Treaty Held with the Ohio Indians at Carlisle in October*, 1753	190
Franklin's Advertisement about Valuation of Wagons for Braddock's Expedition	215
Franklin's Epitaph in His Own Hand	301

These facsimiles from books and manuscripts in the Mason Franklin Collection are reproduced by permission of the Yale University Library.

Introduction

Time alone determines what books are truly classics. Literary critics pronounce one contemporary work a failure, another a masterpiece. More often than not the book that has been praised fades on second reading, the emotional situation that distorted judgment disappears, and succeeding generations marvel at the quaint taste of their grandparents. But there are some books that never lose their appeal. Down through the centuries they speak to men with the same authentic force. They are as vivid and interesting today as they ever were, and they will continue to be so until humanity itself has changed. For the true classics know neither time nor place; they are always modern.

Benjamin Franklin's *Autobiography* is one of these. It has the perennial charm of romance with the added attraction of being absolutely true. It is the first American success story. Beginning as a printer's apprentice, without friends or influence, Franklin accumulated enough wealth to retire soon after he was forty. The forty-odd years that remained to him were devoted entirely to public service. He became our most skillful diplomat, represented the Colonies in England before the Revolution, helped draft the Declaration of Independence, and secured the support of France for the newly founded Republic. He was the first tradesman to hold high public office in America. With only two years of formal schooling, he became our first great literary figure and one of the foremost scientists of his age, whom universities and learned societies on both sides of the Atlantic delighted to honor. When he sat down to write his *Autobiography*, he was quite justified in assuming that others would like to know how this eminence was achieved.

Although his book has been translated into almost every language and is a favorite throughout the world, Franklin had those qualities that we consider peculiarly American. He had the natural adaptability that made him perfectly at home everywhere, setting type with his own hands or testifying before the House of Commons, leading militia against Indians on the frontier or chatting with high-born ladies at the French court. He had also a remarkable endowment of Yankee ingenuity, which he applied to the improvement of everything about him. He invented the Franklin stove to warm houses and the lightning rod to protect them, a combined chair and ladder to reach books and bifocal glasses for reading them. He organized the first police department in Philadelphia and the first public

Introduction

library in America. Under his direction the post office first became profitable and efficient.

He constantly surprises you by anticipating modern discoveries. When he saw the first balloon, his mind leaped at once to the possibility of air mail and parachute troops. "Convincing sovereigns of the folly of wars may perhaps be one effect of it," he wrote to a friend, "since it will be impracticable for the most potent of them to guard his dominions. Five thousand balloons, capable of raising two men each, could not cost more than five ships of the line; and where is the prince who can afford so to cover his country with troops for its defense as that ten thousand men descending from the clouds might not in many places do an infinite deal of mischief before a force could be brought together to repel them?"

With the patience that made his scientific observations so valuable Franklin applied himself to the study of politics. Men, he believed, if one understood their motives, could be managed as easily as smoky chimneys. The persuasive methods that got him cooperation in establishing a hospital or an academy were now turned to the service of the Colony and Nation. He was not an orator and rarely spoke in public. Modestly, with wit and humanity to reinforce his logic, he convinced others that his proposal was directed towards their own best interest.

Kings and Parliamentary committees succumbed to the common sense of his arguments. But the person he found hardest to manage was himself. In this case too he set about the task scientifically, drawing up a list of the virtues he considered desirable and carefully recording his improvement. Temperance in eating and drinking and the avoidance of trivial conversation headed his list of twelve virtues, which tends to treat morality as a mere matter of prudence and worldly wisdom. As an afterthought, at the suggestion of a Quaker friend who said that he was generally thought proud, Franklin added a thirteenth: "Humility. Imitate Jesus and Socrates."

His attitude towards marriage illustrates the latitude of his principles. When Deborah Read became his wife, there was no proof that her first husband was not still alive, though Franklin seemed less concerned about bigamy than the possibility of having to pay the man's debts. Within a year he brought home his illegitimate son William Franklin, whose mother has never been identified. Thirty years later he took charge of William's illegitimate son William Temple Franklin, who became his secretary and constant companion. That such indifference to convention in no wise diminished the universal admiration felt for Franklin shows how completely he was a man of his world.

Of the next world he had little to say. Early in life he shook off the Puritanism of his fathers. At nineteen he wrote and printed in London his *Dissertation on Liberty and Necessity, Pleasure and Pain,* which denied Free Will and doubted the immortality of the soul. As he grew older he became more tolerant. When the Constitutional Convention had been meeting for a month, it was he who first proposed that all future sessions should open with a prayer. Addressing George Washington, who was presiding, Franklin said: "I have lived, Sir, a long time, and the longer I live the more convincing proofs I see of this truth: that God governs in the affairs of men. And if a sparrow cannot fall to the ground without His notice, is it probable that an empire can rise without His aid?" Though skeptical of them all, he contributed to the support of every religion. But his home was on this earth; and the calm intelligence that had robbed the thunderbolt of its terror could await the future with unruffled serenity.

In every field Franklin left his mark. Everywhere human life was richer because he had lived. With the simplest materials—a kite, a silk handkerchief, an old key—he solved mysteries that had puzzled men always. His cool common sense and ability to compromise helped lay the foundations of our national life. The methods by which his success was attained he made available to everyone in this volume.

The Classics Club Text

THE MANUSCRIPT of Franklin's *Autobiography* consists of four separate parts. In the first part, written in 1771 for the information of his son William (Governor of New Jersey, 1762-1776), Franklin brought the story of his life only to the year 1730. At Passy in 1784 the short second section was written, devoted largely to describing the plan for moral improvement. Most of the remainder Franklin composed at Philadelphia in 1788, when he found that he could dictate to his grandson even when he felt too ill to hold a pen. Shortly before his death he wrote a brief final passage, telling of the dispute between the Assembly and the Proprietors of the Colony.

The last event that Franklin mentions occurred in 1760, when his most illustrious career was just beginning. Had he continued, we might read his own story of the Stamp Act and the Boston Tea Party; we might hear a report of his conversations with Edmund Burke or James Boswell or

The Classics Club Text

Louis XVI; we might learn at first hand of his part in drawing up the Declaration of Independence, persuading the French government to support the American Revolution, and making the treaty of peace. To give readers of the Classics Club Edition an account of these years Dr. Alan V. McGee has prepared a continuation of the *Autobiography*, which carries the story from 1760 to Franklin's death thirty years later. Quoting freely from his private letters and other writings, Dr. McGee adds a valuable and unusual feature to this edition.

The history of the publication of the *Autobiography* is exceedingly complex. In 1791 an unauthorized French translation appeared in Paris, made from a revised copy that Franklin had sent to a friend in 1789. The first authentic English edition was published by William Temple Franklin in 1818; but unfortunately he printed it from the 1789 copy, in exchange for which he had given the original manuscript in Franklin's own hand, because it was harder to read! It was not until 1866 that the American Minister to France, John Bigelow, recovered the manuscript and published the original version, which all subsequent texts have followed.

From his study of this manuscript, now at the Huntington Library, Professor Max Farrand has shown that Bigelow was very careless in transcribing, particularly in the matter of punctuation, capital letters, and spelling. Franklin used the customary eighteenth-century forms like *Publick*, *carry'd*, and *Waggon*, beginning every noun with a capital letter; sometimes Bigelow has retained the original, but in thousands of cases he has not.

Accordingly, it has seemed unnecessary to preserve any of these archaic forms. Only the antiquarian or student of language cares about such details. The modern reader wants an accurate text that will give him Franklin's ideas clearly at first glance, without any fumbling of the dictionary. With this end in view the Classics Club Edition provides notes to explain difficult words, and for the first time, it is thought, in an unexpurgated edition, the spelling and punctuation have been modernized throughout. Since Franklin devised our first phonetic alphabet and a remarkable plan for simplified spelling, he would probably approve this change.

Printer

1706-1730

Printer

Twyford, at the Bishop of St. Asaph's, 1771

DEAR SON: I have ever had pleasure in obtaining any little anecdotes of my ancestors. You may remember the inquiries I made among the remains of my relations when you were with me in England and the journey I undertook for that purpose. Imagining it may be equally agreeable to you to know the circumstances of my life, many of which you are yet unacquainted with, and expecting the enjoyment of a week's uninterrupted leisure in my present country retirement, I sit down to write them for you. To which I have besides some other inducements. Having emerged from the poverty and obscurity in which I was born and bred to a state of affluence and some degree of reputation in the world, and having gone so far through life with a considerable share of felicity, the conducing means I made use of, which with the blessing of God so well succeeded, my posterity may like to know, as they may find some of them suitable to their own situations and therefore fit to be imitated.

That felicity, when I reflected on it, has induced me sometimes to say that were it offered to my choice, I should have no objection to a repetition of the same life from its beginning, only asking the advantages authors have in a second edition to correct some faults of the first. So I might, besides correcting the faults, change some sinister accidents and events of it for others more favorable. But though this were denied, I should still accept the offer. Since such a repetition is not to be expected, the next thing most like living one's life over again seems to be a recollection of that life, and to make that recollection as durable as possible by putting it down in writing.

Hereby, too, I shall indulge the inclination so natural in old men, to be talking of themselves and their own past actions; and I shall indulge it without being tiresome to others, who, through respect to age, might conceive themselves obliged to give me a hearing, since this may be read or not as any one pleases. And, lastly (I may as well confess it, since my denial of it will be believed by nobody), perhaps I shall a good deal gratify my own *vanity*. Indeed, I scarce ever heard or saw the introductory words, "Without vanity I may say," etc., but some vain thing immediately followed. Most people dislike vanity in others, whatever share they have of it themselves; but I give it fair quarter wherever I meet with it, being persuaded that it is often productive of good to the possessor and to others that are within his sphere of action; and therefore, in many cases it would not be altogether absurd if a man were to thank God for his vanity among the other comforts of life.

And now I speak of thanking God, I desire with all humility to acknowledge that I owe the mentioned happiness of my past life to His kind providence, which led me to the means I used and gave them success. My belief of this induces me to *hope*, though I must not *presume*, that the same goodness will still be exercised toward me in continuing that happiness or enabling me to bear a fatal reverse, which I may experience as others have done, the complexion of my future fortune being known to Him only in whose power it is to bless to us even our afflictions.

"*Several Particulars Relating to Our Ancestors . . .*"

THE NOTES one of my uncles (who had the same kind of curiosity in collecting family anecdotes) once put into my hands furnished me with several particulars relating to our ancestors. From these notes I learned that the family[1] had lived in the same village, Ecton, in Northamptonshire, for three hundred years, and how much longer he knew not (perhaps from the time when the

---

[1] It is an interesting coincidence that the ancestors of Franklin and Washington came from the same English county. Washington's forebears were proprietors of Sulgrave.

name of Franklin, that before was the name of an order of people,[1] was assumed by them as a surname when others took surnames all over the kingdom), on a freehold of about thirty acres, aided by the smith's business, which had continued in the family till his time, the eldest son being always bred to that business; a custom which he and my father followed as to their eldest sons. When I searched the registers at Ecton, I found an account of their births, marriages, and burials from the year 1555 only, there being no registers kept in that parish at any time preceding. By that register I perceived that I was the youngest son of the youngest son for five generations back. My grandfather Thomas, who was born in 1598, lived at Ecton till he grew too old to follow business longer, when he went to live with his son John, a dyer at Banbury, in Oxfordshire, with whom my father served an apprenticeship. There my grandfather died and lies buried. We saw his gravestone in 1758. His eldest son Thomas lived in the house at Ecton, and left it with the land to his only child, a daughter, who, with her husband, one Fisher, of Wellingborough, sold it to Mr. Isted, now lord of the manor there. My grandfather had four sons that grew up, viz.: Thomas, John, Benjamin and Josiah. I will give you what account I can of them, at this distance from my papers, and if these are not lost in my absence, you will among them find many more particulars.

[1] Freeholders who owned their land.

Autobiography 7

Thomas was bred a smith under his father; but, being ingenious, and encouraged in learning (as all his[1] brothers were) by an Esquire Palmer, then the principal gentleman in that parish, he qualified himself for the business of scrivener;[2] became a considerable man in the county; was a chief mover of all public-spirited undertakings for the county or town of Northampton, and his own village, of which many instances were related of him; and much taken notice of and patronized by the then Lord Halifax. He died in 1702, January 6, old style,[3] just four years to a day before I was born. The account we received of his life and character from some old people at Ecton, I remember, struck you as something extraordinary, from its similarity to what you knew of mine. "Had he died on the same day," you said, "one might have supposed a transmigration."

John was bred a dyer, I believe of woolens. Benjamin was bred a silk dyer, serving an apprenticeship at London. He was an ingenious man. I remember him well, for when I was a boy he came over to my father in Boston, and lived in the house with us some years. He lived to a great age. His grandson, Samuel Franklin, now lives in Boston. He left behind him two quarto volumes, MS., of his own poetry, consisting of little occasional pieces addressed to

[1] The texts read *my*, an obvious slip.
[2] Professional copyist who drew up contracts, etc.
[3] In 1752 eleven days were dropped from the English calendar; Franklin's birthday then became January 17. He used to celebrate both dates.

his friends and relations, of which the following, sent to me, is a specimen.

ACROSTIC

Sent to Benjamin Franklin in New England, July 15, 1710.

> Be to thy parents an obedient son;
> Each day let duty constantly be done;
> Never give way to sloth, or lust, or pride,
> If free you'd be from thousand ills beside;
> Above all ills be sure avoid the shelf;[1]
> Man's danger lies in Satan, sin, and self.
> In virtue, learning, wisdom, progress make;
> Ne'er shrink at suffering for thy Saviour's sake.
>
> Fraud and all falsehood in thy dealings flee,
> Religious always in thy station be;
> Adore the Maker of thy inward part,
> Now's the accepted time, give him thy heart;
> Keep a good conscience, 'tis a constant friend;
> Like judge and witness this thy acts attend.
> In heart with bended knee, alone, adore
> None but the Three in One for evermore.

He had formed a shorthand of his own, which he taught me, but, never practising it, I have now forgot it. I was named after this uncle, there being a particular affection between him and my father. He was very pious, a great

---

[1] Pawnshop.

Autobiography

attender of sermons of the best preachers, which he took down in his shorthand, and had with him many volumes of them. He was also much of a politician, too much, perhaps, for his station. There fell lately into my hands in London a collection he had made of all the principal pamphlets relating to public affairs from 1641 to 1717; many of the volumes are wanting as appears by the numbering, but there still remain eight volumes in folio and twenty-four in quarto and in octavo. A dealer in old books met with them, and knowing me by my sometimes buying of him, he brought them to me. It seems my uncle must have left them here when he went to America, which was above fifty years since. There are many of his notes in the margins.

This obscure family of ours was early in the Reformation and continued Protestants through the reign of Queen Mary, when they were sometimes in danger of trouble on account of their zeal against popery. They had got an English Bible, and to conceal and secure it, it was fastened open with tapes under and within the cover of a joint-stool.[1] When my great-great-grandfather read it to his family, he turned up the joint-stool upon his knees, turning over the leaves then under the tapes. One of the children stood at the door to give notice if he saw the apparitor[2] coming, who was an officer of the spiritual court. In that case the stool was turned down again upon its feet,

[1] Made by a cabinet-maker instead of homemade.
[2] Summoner to the Church Court.

when the Bible remained concealed under it as before. This anecdote I had from my uncle Benjamin. The family continued all of the Church of England till about the end of Charles the Second's reign, when some of the ministers that had been outed for non-conformity holding conventicles[1] in Northamptonshire, Benjamin and Josiah adhered to them, and so continued all their lives; the rest of the family remained with the Episcopal Church.

"The Youngest Child But Two . . ."

JOSIAH, my father, married young and carried his wife with three children into New England about 1682. The conventicles having been forbidden by law and frequently disturbed, induced some considerable men of his acquaintance to remove to that country, and he was prevailed with to accompany them thither, where they expected to enjoy their mode of religion with freedom. By the same wife he had four children more born there, and by a second wife ten more, in all seventeen; of which I remember thirteen sitting at one time at his table, who all grew up to be men and women and married; I was the youngest son and the youngest child but two, and was born in Boston, New England.[2] My mother, the second wife, was Abiah Folger, daughter of Peter Folger, one of the first

[1] Meetings of dissenters, in this case Presbyterians.
[2] Franklin was born in Milk Street, opposite the Old South Church of which his parents were members and where he was baptized the day of his birth.

settlers of New England, of whom honorable mention is made by Cotton Mather in his church history of that country entitled *Magnalia Christi Americana* as "a godly, learned Englishman," if I remember the words rightly. I have heard that he wrote sundry small occasional pieces, but only one of them was printed, which I saw now many years since. It was written in 1675 in the homespun verse of that time and people, and addressed to those then concerned in the government there. It was in favor of liberty of conscience and in behalf of the Baptists, Quakers, and other sectaries that had been under persecution, ascribing the Indian wars and other distresses that had befallen the country to that persecution as so many judgments of God to punish so heinous an offense, and exhorting a repeal of those uncharitable laws. The whole appeared to me as written with a good deal of decent plainness and manly freedom. The six concluding lines I remember, though I have forgotten the two first of the stanza; but the purport of them was that his censures proceeded from good will and, therefore, he would be known to be the author.

> *Because to be a libeller (says he)*
>   *I hate it with my heart;*
> *From Sherburne*[1] *town, where now I dwell*
>   *My name I do put here;*
> *Without offence your real friend,*
>   *It is Peter Folgier.*

[1] Now Nantucket.

My elder brothers were all put apprentices to different trades. I was put to the grammar-school [1] at eight years of age, my father intending to devote me, as the tithe [2] of his sons, to the service of the Church. My early readiness in learning to read (which must have been very early, as I do not remember when I could not read) and the opinion of all his friends that I should certainly make a good scholar encouraged him in this purpose of his. My uncle Benjamin, too, approved of it, and proposed to give me all his shorthand volumes of sermons, I suppose as a stock to set up with, if I would learn his character.[3] I continued, however, at the grammar-school not quite one year, though in that time I had risen gradually from the middle of the class of that year to be the head of it, and farther was removed into the next class above it, in order to go with that into the third at the end of the year. But my father in the meantime from a view of the expense of a college education, which having so large a family he could not well afford, and the mean living many so educated were afterwards able to obtain—reasons that he gave to his friends in my hearing—altered his first intention, took me from the grammar-school, and sent me to a school for writing and arithmetic kept by a then famous man, Mr. George Brownell, very successful in his profession generally, and that by mild, encouraging methods. Under him

[1] Now the Boston Latin School.
[2] The tenth set aside for the Church.
[3] Handwriting.

Autobiography

I acquired fair writing pretty soon, but I failed in the arithmetic and made no progress in it. At ten years old I was taken home to assist my father in his business, which was that of a tallow-chandler and soap-boiler, a business he was not bred to, but had assumed on his arrival in New England and on finding his dyeing trade would not maintain his family, being in little request. Accordingly, I was employed in cutting wick for the candles, filling the dipping mould and the moulds for cast candles, attending the shop, going of errands, etc.

I disliked the trade and had a strong inclination for the sea, but my father declared against it; however, living near the water, I was much in and about it, learned early to swim well, and to manage boats; and when in a boat or canoe with other boys, I was commonly allowed to govern, especially in any case of difficulty; and upon other occasions I was generally a leader among the boys, and sometimes led them into scrapes, of which I will mention one instance, as it shows an early projecting public spirit, though not then justly conducted.

There was a salt-marsh that bounded part of the mill-pond, on the edge of which at high water we used to stand to fish for minnows. By much trampling we had made it a mere quagmire. My proposal was to build a wharf there fit for us to stand upon, and I showed my comrades a large heap of stones which were intended for a new house near the marsh and which would very well suit our purpose. Accordingly, in the evening when the workmen were gone, I assembled a number of my playfellows and, working with them diligently like so many emmets,[1] sometimes two or three to a stone, we brought them all away and built our little wharf. The next morning the workmen were surprised at missing the stones, which were found in our wharf. Inquiry was made after the removers; we were discovered and complained of; several of us were corrected by our fathers; and, though I pleaded the usefulness of the work, mine convinced me that nothing was useful which was not honest.

I think you may like to know something of his person and character. He had an excellent constitution of body, was of middle stature, but well set and very strong; he was ingenious, could draw prettily, was skilled a little in music, and had a clear pleasing voice, so that when he played psalm tunes on his violin and sung withal, as he sometimes did in an evening after the business of the day was over, it was extremely agreeable to hear. He had a mechanical genius too, and on occasion was very handy in the use of other tradesmen's tools; but his great excellence lay in a sound understanding and solid judgment in prudential[2] matters both in private and public affairs. In the latter, indeed, he was never employed, the numerous family he had to educate and the straitness[3] of his circumstances keeping him close to his trade; but I remember well his being frequently visited by leading peo-

---

[1] Ants.   [2] Practical.   [3] Limited resources.

ple, who consulted him for his opinion in affairs of the town or of the church he belonged to and showed a good deal of respect for his judgment and advice: he was also much consulted by private persons about their affairs when any difficulty occurred and frequently chosen an arbitrator between contending parties. At his table he liked to have as often as he could some sensible friend or neighbor to converse with, and always took care to start some ingenious or useful topic for discourse, which might tend to improve the minds of his children. By this means he turned our attention to what was good, just, and prudent in the conduct of life; and little or no notice was ever taken of what related to the victuals on the table, whether it was well or ill dressed, in or out of season, of good or bad flavor, preferable or inferior to this or that other thing of the kind, so that I was brought up in such a perfect inattention to those matters as to be quite indifferent what kind of food was set before me, and so unobservant of it that to this day if I am asked I can scarce tell a few hours after dinner what I dined upon. This has been a convenience to me in traveling, where my companions have been sometimes very unhappy for want of a suitable gratification of their more delicate, because better instructed, tastes and appetites.

My mother had likewise an excellent constitution: she suckled all her ten children. I never knew either my father or mother to have any sickness but that of which they died, he at eighty-nine, and she at eighty-five years of age. They lie buried together at Boston, where I some years since placed a marble over their grave, with this inscription:

JOSIAH FRANKLIN,
and
ABIAH his wife,
lie here interred.
They lived lovingly together in wedlock
fifty-five years.
Without an estate, or any gainful employment,
By constant labour and industry,
with God's blessing,
They maintained a large family
comfortably,
and brought up thirteen children
and seven grandchildren
reputably.
From this instance, reader,
Be encouraged to diligence in thy calling,
And distrust not Providence.
He was a pious and prudent man;
She, a discreet and virtuous woman.
Their youngest son,
In filial regard to their memory,
Places this stone.
J. F. born 1655, died 1744, Ætat 89.
A. F. born 1667, died 1752, —— 85.

By my rambling digressions I perceive myself to be grown old. I used to write more methodically. But one does not dress for private company as for a public ball. 'Tis perhaps only negligence.

To return: I continued thus employed in my father's business for two years, that is, till I was twelve years old; and my brother John, who was bred to that business, having left my father, married, and set up for himself at Rhode Island, there was all appearance that I was destined to supply his place, and become a tallow-chandler. But my dislike to the trade continuing, my father was under apprehensions that if he did not find one for me more agreeable, I should break away and get to sea as his son Josiah had done, to his great vexation. He therefore sometimes took me to walk with him and see joiners, bricklayers, turners, braziers,[1] etc., at their work, that he might observe my inclination, and endeavor to fix it on some trade or other on land. It has ever since been a pleasure to me to see good workmen handle their tools; and it has been useful to me, having learned so much by it as to be able to do little jobs myself in my house when a workman could not readily be got, and to construct little machines for my experiments while the intention of making the experiment was fresh and warm in my mind. My father at last fixed upon the cutler's trade, and my uncle Benjamin's son Samuel, who was bred to that business in London, being about that time established in Boston, I

[1] Brass workers.

was sent to be with him some time on liking. But his expectations of a fee with me displeasing my father, I was taken home again.

"Bound to My Brother . . ."

FROM a child I was fond of reading, and all the little money that came into my hands was ever laid out in books. Pleased with the *Pilgrim's Progress* my first collection was of John Bunyan's works in separate little volumes. I afterward sold them to enable me to buy R. Burton's *Historical Collections;* they were small chapmen's books,[1] and cheap, 40 or 50 in all. My father's little library consisted chiefly of books in polemic divinity,[2] most of which I read, and have since often regretted that, at a time when I had such a thirst for knowledge, more proper books had not fallen in my way, since it was now resolved I should not be a clergyman. *Plutarch's Lives* there was in which I read abundantly, and I still think that time spent to great advantage. There was also a book of Defoe's called an *Essay on Projects* and another of Dr. Mather's called *Essays to do Good*, which perhaps gave me a turn of thinking that had an influence on some of the principal future events of my life.

This bookish inclination at length determined my father to make me a printer, though he had already one son

[1] Small paper-bound books sold by peddlers.
[2] Controversial religious books.

Autobiography

(James) of that profession. In 1717 my brother James returned from England with a press and letters to set up his business in Boston. I liked it much better than that of my father, but still had a hankering for the sea. To prevent the apprehended effect of such an inclination my father was impatient to have me bound [1] to my brother. I stood out some time, but at last was persuaded, and signed the indentures [2] when I was yet but twelve years old. I was to serve as an apprentice till I was twenty-one years of age, only I was to be allowed journeyman's [3] wages during the last year. In a little time I made great proficiency in the business, and became a useful hand to my brother. I now had access to better books. An acquaintance with the apprentices of booksellers enabled me sometimes to borrow a small one, which I was careful to return soon and clean. Often I sat up in my room reading the greatest part of the night, when the book was borrowed in the evening and to be returned early in the morning, lest it should be missed or wanted.

And after some time an ingenious tradesman, Mr. Matthew Adams, who had a pretty collection of books and who frequented our printing-house, took notice of me, invited me to his library, and very kindly lent me such books as I chose to read. I now took a fancy to poetry and made some little pieces; my brother, thinking it might

[1] Apprenticed. [2] Contracts.
[3] Independent worker, hired by the day.

turn to account, encouraged me and put me on composing occasional ballads. One was called *The Lighthouse Tragedy*, and contained an account of the drowning of Captain Worthilake with his two daughters; the other was a sailor's song on the taking of Teach (or Blackbeard) the pirate. They were wretched stuff in the Grub-Street-ballad style, and when they were printed he sent me about the town to sell them. The first sold wonderfully, the event being recent, having made a great noise. This flattered my vanity; but my father discouraged me by ridiculing my performances and telling me verse-makers were generally beggars. So I escaped being a poet, most probably a very bad one; but as prose writing has been of great use to me in the course of my life and was a principal means of my advancement, I shall tell you how in such a situation I acquired what little ability I have in that way.

There was another bookish lad in the town, John Collins by name, with whom I was intimately acquainted. We sometimes disputed, and very fond we were of argument and very desirous of confuting one another, which disputatious turn, by the way, is apt to become a very bad habit, making people often extremely disagreeable in company by the contradiction that is necessary to bring it into practice; and thence, besides souring and spoiling the conversation, is productive of disgusts and perhaps enmities where you may have occasion for friendship. I had caught it by reading my father's books of dispute

Autobiography

about religion. Persons of good sense, I have since observed, seldom fall into it except lawyers, university men, and men of all sorts that have been bred at Edinburgh.

A question was once, somehow or other, started between Collins and me of the propriety of educating the female sex in learning and their abilities for study. He was of opinion that it was improper and that they were naturally unequal to it. I took the contrary side, perhaps a little for dispute's sake. He was naturally more eloquent, had a ready plenty of words, and sometimes, as I thought, bore me down more by his fluency than by the strength of his reasons. As we parted without settling the point and were not to see one another again for some time, I sat down to put my arguments in writing, which I copied fair and sent to him. He answered, and I replied. Three or four letters of a side had passed when my father happened to find my papers and read them. Without entering into the discussion he took occasion to talk to me about the manner of my writing; observed that, though I had the advantage of my antagonist in correct spelling and pointing[1] (which I owed to the printing-house), I fell far short in elegance of expression, in method, and in perspicuity, of which he convinced me by several instances. I saw the justice of his remarks and thence grew more attentive to the manner in writing and determined to endeavor at improvement.

About this time I met with an odd volume of the *Spec-*

[1] Punctuation.

tator.[1] It was the third. I had never before seen any of them. I bought it, read it over and over, and was much delighted with it. I thought the writing excellent and wished, if possible, to imitate it. With this view I took some of the papers and, making short hints of the sentiment in each sentence, laid them by a few days, and then without looking at the book tried to complete the papers again by expressing each hinted sentiment at length and as fully as it had been expressed before in any suitable words that should come to hand. Then I compared my *Spectator* with the original, discovered some of my faults, and corrected them. But I found I wanted a stock of words or a readiness in recollecting and using them, which I thought I should have acquired before that time if I had gone on making verses; since the continual occasion for words of the same import but of different length to suit the measure, or of different sound for the rhyme, would have laid me under a constant necessity of searching for variety and also have tended to fix that variety in my mind and make me master of it. Therefore I took some of the tales and turned them into verse and after a time, when I had pretty well forgotten the prose, turned them back again. I also sometimes jumbled my collections of hints into confusion and after some weeks endeavored to reduce them into the best order before I began to form the full sentences and complete the paper. This was to teach me

[1] An English periodical famous for contributions by Joseph Addison and Richard Steele.

method in the arrangement of thoughts. By comparing my work afterwards with the original I discovered many faults and amended them; but I sometimes had the pleasure of fancying that in certain particulars of small import I had been lucky enough to improve the method or the language, and this encouraged me to think I might possibly in time come to be a tolerable English writer, of which I was extremely ambitious. My time for these exercises and for reading was at night, after work or before it began in the morning, or on Sundays, when I contrived to be in the printing-house alone, evading as much as I could the common attendance on public worship which my father used to exact of me when I was under his care, and which indeed I still thought a duty, though I could not, as it seemed to me, afford time to practise it.

"I Happened to Meet with a Book . . ."

WHEN about sixteen years of age I happened to meet with a book, written by one Tryon, recommending a vegetable diet. I determined to go into it. My brother, being yet unmarried, did not keep house, but boarded himself and his apprentices in another family. My refusing to eat flesh occasioned an inconveniency, and I was frequently chid for my singularity.[1] I made myself acquainted with Tryon's manner of preparing some of his dishes, such as boiling potatoes or rice, making hasty pud-

[1] Scolded for being different.

ding, and a few others, and then proposed to my brother that if he would give me, weekly, half the money he paid for my board, I would board myself. He instantly agreed to it, and I presently found that I could save half what he paid me. This was an additional fund for buying books. But I had another advantage in it. My brother and the rest going from the printing-house to their meals, I remained there alone and dispatching presently my light repast, which often was no more than a biscuit or a slice of bread, a handful of raisins, or a tart from the pastry-cook's, and a glass of water, had the rest of the time till their return for study, in which I made the greater progress from that greater clearness of head and quicker apprehension which usually attend temperance in eating and drinking.

And now it was that, being on some occasion made ashamed of my ignorance in figures, which I had twice failed in learning when at school, I took Cocker's book of arithmetic and went through the whole by myself with great ease. I also read Seller's and Shermy's books of navigation and became acquainted with the little geometry they contain, but never proceeded far in that science. And I read about this time Locke *On Human Understanding* and the *Art of Thinking* by Messrs. du Port Royal.

While I was intent on improving my language, I met with an English grammar (I think it was Greenwood's) at the end of which there were two little sketches of the

Autobiography

arts of rhetoric and logic, the latter finishing with a specimen of a dispute in the Socratic method; [1] and soon after I procured Xenophon's *Memorable Things of Socrates*, wherein there are many instances of the same method. I was charmed with it, adopted it, dropped my abrupt contradiction and positive argumentation, and put on the humble inquirer and doubter. And being then from reading Shaftesbury and Collins become a real doubter in many points of our religious doctrine, I found this method safest for myself and very embarrassing to those against whom I used it; therefore I took a delight in it, practised it continually, and grew very artful and expert in drawing people, even of superior knowledge, into concessions the consequences of which they did not foresee, entangling them in difficulties out of which they could not extricate themselves and so obtaining victories that neither myself nor my cause always deserved. I continued this method some few years, but gradually left it, retaining only the habit of expressing myself in terms of modest diffidence; never using, when I advanced anything that may possibly be disputed, the words *certainly*, *undoubtedly*, or any others that give the air of positiveness to an opinion; but rather say, "I conceive" or "apprehend" a thing to be so and so; "it appears to me," or "I should think it so or so," for such and such reasons; or "I imagine it to be so"; or "it is so, if I am not mistaken." This habit, I

---

[1] By asking apparently innocent questions Socrates led his opponent to admit the truth of his position.

believe, has been of great advantage to me when I have had occasion to inculcate my opinions and persuade men into measures that I have been from time to time engaged in promoting; and, as the chief ends of conversation are to *inform* or to *be informed*, to *please* or to *persuade*, I wish well-meaning, sensible men would not lessen their power of doing good by a positive, assuming manner, that seldom fails to disgust, tends to create opposition and to defeat every one of those purposes for which speech was given to us, to wit, giving or receiving information or pleasure. For, if you would inform, a positive and dogmatical manner in advancing your sentiments may provoke contradiction and prevent a candid attention. If you wish information and improvement from the knowledge of others, and yet at the same time express yourself as firmly fixed in your present opinions, modest, sensible men, who do not love disputation, will probably leave you undisturbed in the possession of your error. And by such a manner you can seldom hope to recommend yourself in *pleasing* your hearers or to persuade those whose concurrence you desire. Pope says, judiciously:

> *Men should be taught as if you taught them not,*
> *And things unknown propos'd as things forgot;*

farther recommending to us "To speak, tho' sure, with seeming diffidence." And he might have coupled with this line that which he has coupled with another, I think, less properly, "For want of modesty is want of sense." If you ask, Why less properly? I must repeat the lines,

> *Immodest words admit of no defense,*
> *For want of modesty is want of sense.*

Now, is not *want of sense* (where a man is so unfortunate as to want it) some apology for his *want of modesty?* and would not the lines stand more justly thus?

> *Immodest words admit but this defense,*
> *That want of modesty is want of sense.*

This, however, I should submit to better judgments.

"Writing Little Pieces for This Paper . . ."

MY BROTHER had in 1720 or 1721 begun to print a newspaper. It was the second that appeared in America and was called the *New England Courant*. The only one before it was the *Boston News-Letter*. I remember his being dissuaded by some of his friends from the undertaking as not likely to succeed, one newspaper being, in their judgment, enough for America. At this time (1771) there are not less than five-and-twenty. He went on, however, with the undertaking; and after having worked in composing the types and printing off the sheets, I was employed to carry the papers through the streets to the customers.

He had some ingenious men among his friends who

amused themselves by writing little pieces for this paper which gained it credit and made it more in demand, and these gentlemen often visited us. Hearing their conversations and their accounts of the approbation their papers were received with, I was excited to try my hand among them; but, being still a boy and suspecting that my brother would object to printing anything of mine in his paper if he knew it to be mine, I contrived to disguise my hand, and, writing an anonymous paper, I put it in at night under the door of the printing-house. It was found in the morning and communicated to his writing friends when they called in as usual. They read it, commented on it in my hearing, and I had the exquisite pleasure of finding it met with their approbation and that in their different guesses at the author none were named but men of some character among us for learning and ingenuity. I suppose now that I was rather lucky in my judges and that perhaps they were not really so very good ones as I then esteemed them.

Encouraged, however, by this, I wrote and conveyed in the same way to the press several more papers which were equally approved; and I kept my secret till my small fund of sense for such performances was pretty well exhausted, and then I discovered [1] it, when I began to be considered a little more by my brother's acquaintance and in a manner that did not quite please him, as he thought.

[1] Revealed.

Autobiography

probably with reason, that it tended to make me too vain. And perhaps this might be one occasion of the differences that we began to have about this time. Though a brother, he considered himself as my master and me as his apprentice and accordingly expected the same services from me as he would from another, while I thought he demeaned me too much in some he required of me, who from a brother expected more indulgence. Our disputes were often brought before our father, and I fancy I was either generally in the right or else a better pleader, because the judgment was generally in my favor. But my brother was passionate [1] and had often beaten me, which I took extremely amiss; and, thinking my apprenticeship very tedious, I was continually wishing for some opportunity of shortening it, which at length offered in a manner unexpected.[2]

One of the pieces in our newspaper on some political point, which I have now forgotten, gave offense to the Assembly. He was taken up, censured, and imprisoned for a month by the speaker's warrant, I suppose because he would not discover his author. I, too, was taken up and examined before the Council; but, though I did not give them any satisfaction, they contented themselves with admonishing me and dismissed me, considering me,

[1] Hot-tempered.
[2] I fancy his harsh and tyrannical treatment of me might be a means of impressing me with that aversion to arbitrary power that has stuck to me through my whole life.—FRANKLIN.

perhaps, as an apprentice who was bound to keep his master's secrets.

During my brother's confinement, which I resented a good deal, notwithstanding our private differences, I had the management of the paper; and I made bold to give our rulers some rubs in it, which my brother took very kindly, while others began to consider me in an unfavorable light as a young genius that had a turn for libeling and satire. My brother's discharge was accompanied with an order of the House (a very odd one), that "James Franklin should no longer print the paper called the *New England Courant*."

There was a consultation held in our printing-house among his friends, what he should do in this case. Some proposed to evade the order by changing the name of the paper; but my brother, seeing inconveniences in that, it was finally concluded on as a better way to let it be printed for the future under the name of BENJAMIN FRANKLIN; and to avoid the censure of the Assembly, that might fall on him as still printing it by his apprentice, the contrivance was that my old indenture should be returned to me with a full discharge on the back of it to be shown on occasion, but to secure to him the benefit of my service I was to sign new indentures for the remainder of the term which were to be kept private. A very flimsy scheme it was; however, it was immediately executed, and the paper went on accordingly under my name for several months.

"I Took Upon Me to Assert My Freedom . . ."

AT LENGTH, a fresh difference arising between my brother and me, I took upon me to assert my freedom, presuming that he would not venture to produce the new indentures. It was not fair in me to take this advantage, and this I therefore reckon one of the first errata [1] of my life; but the unfairness of it weighed little with me when under the impressions of resentment for the blows his passion too often urged him to bestow upon me, though he was otherwise not an ill-natured man: perhaps I was too saucy and provoking.

When he found I would leave him, he took care to prevent my getting employment in any other printing-house of the town by going round and speaking to every master, who accordingly refused to give me work. I then thought of going to New York as the nearest place where there was a printer; and I was rather inclined to leave Boston when I reflected that I had already made myself a little obnoxious to the governing party, and from the arbitrary proceeding of the Assembly in my brother's case it was likely I might if I stayed soon bring myself into scrapes; and farther, that my indiscreet disputations about religion began to make me pointed at with horror by good people as an infidel or atheist. I determined on the point, but my

[1] Mistakes.

father now siding with my brother, I was sensible that if I attempted to go openly, means would be used to prevent me. My friend Collins, therefore, undertook to manage a little for me. He agreed with the captain of a New York sloop for my passage under the notion of my being a young acquaintance of his that had got a naughty girl with child, whose friends would compel me to marry her and therefore I could not appear or come away publicly. So I sold some of my books to raise a little money, was taken on board privately, and as we had a fair wind, in three days I found myself in New York, near 300 miles from home, a boy of but seventeen, without the least recommendation to or knowledge of any person in the place, and with very little money in my pocket.

My inclinations for the sea were by this time worn out, or I might now have gratified them. But, having a trade and supposing myself a pretty good workman, I offered my service to the printer in the place, old Mr. William Bradford, who had been the first printer in Pennsylvania but removed from thence upon the quarrel of George Keith. He could give me no employment, having little to do, and help enough already; but says he, "My son at Philadelphia has lately lost his principal hand, Aquila Rose, by death; if you go thither I believe he may employ you." Philadelphia was a hundred miles further; I set out, however, in a boat for Amboy, leaving my chest and things to follow me round by sea.

In crossing the bay we met with a squall that tore our

Autobiography

rotten sails to pieces, prevented our getting into the Kill,[1] and drove us upon Long Island. In our way a drunken Dutchman, who was a passenger too, fell overboard; when he was sinking I reached through the water to his shock pate,[2] and drew him up, so that we got him in again. His ducking sobered him a little and he went to sleep, taking first out of his pocket a book which he desired I would dry for him. It proved to be my old favorite author Bunyan's *Pilgrim's Progress* in Dutch, finely printed on good paper with copper cuts, a dress better than I had ever seen it wear in its own language. I have since found that it has been translated into most of the languages of Europe and suppose it has been more generally read than any other book except perhaps the Bible. Honest John was the first that I know of who mixed narration and dialogue; a method of writing very engaging to the reader, who in the most interesting parts finds himself, as it were, brought into the company and present at the discourse. Defoe in his *Crusoe*, his *Moll Flanders, Religious Courtship, Family Instructor*, and other pieces, has imitated it with success; and Richardson has done the same in his *Pamela*, etc.

When we drew near the island, we found it was at a place where there could be no landing, there being a great surf on the stony beach. So we dropped anchor and swung round towards the shore. Some people came down to the

---

[1] Creek or channel west of Staten Island on the inside route to Perth Amboy.
[2] Shaggy head.

water edge and hallooed to us, as we did to them; but the wind was so high and the surf so loud that we could not hear so as to understand each other. There were canoes on the shore, and we made signs and hallooed that they should fetch us; but they either did not understand us or thought it impracticable. So they went away, and night coming on, we had no remedy but to wait till the wind should abate; and, in the meantime, the boatman and I concluded to sleep if we could; and so crowded into the scuttle [1] with the Dutchman, who was still wet, and the spray beating over the head of our boat leaked through to us so that we were soon almost as wet as he. In this manner we lay all night with very little rest; but, the wind abating the next day, we made a shift to reach Amboy before night, having been thirty hours on the water without victuals or any drink but a bottle of filthy rum, and the water we sailed on being salt.

In the evening I found myself very feverish and went into bed; but, having read somewhere that cold water drunk plentifully was good for a fever, I followed the prescription, sweat plentifully most of the night, my fever left me, and in the morning, crossing the ferry, I proceeded on my journey on foot, having fifty miles to Burlington, where I was told I should find boats that would carry me the rest of the way to Philadelphia.

It rained very hard all the day; I was thoroughly soaked, and by noon a good deal tired; so I stopped at a poor inn,

[1] Covered opening in the deck.

Autobiography

where I stayed all night, beginning now to wish that I had never left home. I cut so miserable a figure, too, that I found by the questions asked me I was suspected to be some runaway servant and in danger of being taken up on that suspicion. However, I proceeded the next day and got in the evening to an inn within eight or ten miles of Burlington, kept by one Dr. Brown. He entered into conversation with me while I took some refreshment, and, finding I had read a little, became very sociable and friendly. Our acquaintance continued as long as he lived. He had been, I imagine, an itinerant doctor, for there was no town in England or country in Europe of which he could not give a very particular account. He had some letters and was ingenious,[1] but much of an unbeliever, and wickedly undertook some years after to travesty the Bible in doggerel verse as Cotton had done Virgil. By this means he set many of the facts in a very ridiculous light and might have hurt weak minds if his work had been published, but it never was.

At his house I lay that night and the next morning reached Burlington, but had the mortification to find that the regular boats were gone a little before my coming and no other expected to go before Tuesday, this being Saturday; wherefore I returned to an old woman in the town of whom I had bought gingerbread to eat on the water, and asked her advice. She invited me to lodge at her house till a passage by water should offer; and being

[1] Some education and was clever.

tired with my foot traveling, I accepted the invitation. She, understanding I was a printer, would have had me stay at that town and follow my business, being ignorant of the stock necessary to begin with. She was very hospitable, gave me a dinner of ox-cheek with great good will, accepting only of a pot of ale in return; and I thought myself fixed till Tuesday should come. However, walking in the evening by the side of the river, a boat came by, which I found was going towards Philadelphia with several people in her. They took me in, and, as there was no wind, we rowed all the way; and about midnight, not having yet seen the city, some of the company were confident we must have passed it, and would row no farther; the others knew not where we were; so we put toward the shore, got into a creek, landed near an old fence, with the rails of which we made a fire, the night being cold, in October, and there we remained till daylight. Then one of the company knew the place to be Cooper's Creek, a little above Philadelphia, which we saw as soon as we got out of the creek, and arrived there about eight or nine o'clock on the Sunday morning and landed at the Market Street wharf.[1]

"Such Unlikely Beginnings . . ."

I HAVE been the more particular in this description of my journey and shall be so of my first entry into that city

[1] Philadelphia was at this time (1723) a city of 7,000 inhabitants.

that you may in your mind compare such unlikely beginnings with the figure I have since made there. I was in my working dress, my best clothes being to come round by sea. I was dirty from my journey; my pockets were stuffed out with shirts and stockings, and I knew no soul nor where to look for lodging. I was fatigued with traveling, rowing, and want of rest; I was very hungry; and my whole stock of cash consisted of a Dutch dollar and about a shilling in copper. The latter I gave the people of the boat for my passage, who at first refused it, on account of my rowing; but I insisted on their taking it. A man being sometimes more generous when he has but a little money than when he has plenty, perhaps through fear of being thought to have but little.

Then I walked up the street, gazing about till near the market-house I met a boy with bread. I had made many a meal on bread, and, inquiring where he got it, I went immediately to the baker's he directed me to in Second Street and asked for biscuit, intending such as we had in Boston; but they, it seems, were not made in Philadelphia. Then I asked for a three-penny loaf, and was told they had none such. So not considering or knowing the difference of money and the greater cheapness nor the names of his bread, I bade him give me three-penny worth of any sort. He gave me, accordingly, three great puffy rolls. I was surprised at the quantity, but took it, and, having no room in my pockets, walked off with a roll under each arm and eating the other. Thus I went up Market Street as far as Fourth Street, passing by the door of Mr. Read, my future wife's father; when she,[1] standing at the door, saw me, and thought I made, as I certainly did, a most awkward, ridiculous appearance. Then I turned and went down Chestnut Street and part of Walnut Street, eating my roll all the way, and coming round found myself again at Market Street wharf near the boat I came in, to which I went for a draught of the river water; and, being filled with one of my rolls, gave the other two to a woman and her child that came down the river in the boat with us and were waiting to go farther.

Thus refreshed I walked again up the street, which by this time had many clean-dressed people in it, who were all walking the same way. I joined them and thereby was led into the great meeting-house of the Quakers near the market. I sat down among them and after looking round awhile and hearing nothing said, being very drowsy through labor and want of rest the preceding night, I fell fast asleep, and continued so till the meeting broke up, when one was kind enough to rouse me. This was, therefore, the first house I was in or slept in, in Philadelphia.

Walking down again toward the river and looking in the faces of people, I met a young Quaker man whose countenance I liked, and, accosting him, requested he would tell me where a stranger could get lodging. We were then near the sign of the Three Mariners. "Here," says he, "is one place that entertains strangers, but it is

[1] Deborah Read.

not a reputable house; if thee wilt walk with me, I'll show thee a better." He brought me to the Crooked Billet in Water Street. Here I got a dinner; and while I was eating it several sly questions were asked me, as it seemed to be suspected from my youth and appearance that I might be some runaway.

After dinner my sleepiness returned, and being shown to a bed, I lay down without undressing and slept till six in the evening, was called to supper, went to bed again very early, and slept soundly till next morning. Then I made myself as tidy as I could and went to Andrew Bradford the printer's. I found in the shop the old man his father, whom I had seen at New York, and who, traveling on horseback, had got to Philadelphia before me. He introduced me to his son, who received me civilly, gave me a breakfast, but told me he did not at present want a hand, being lately supplied with one; but there was another printer in town, lately set up, one Keimer, who, perhaps, might employ me; if not, I should be welcome to lodge at his house, and he would give me a little work to do now and then till fuller business should offer.

"*A Crafty Old Sophister and a Novice . . .*"

THE OLD gentleman said he would go with me to the new printer; and when we found him, "Neighbor," says Bradford, "I have brought to see you a young man of your business; perhaps you may want such a one." He asked me a few questions, put a composing stick [1] in my hand to see how I worked, and then said he would employ me soon, though he had just then nothing for me to do; and, taking old Bradford, whom he had never seen before, to be one of the town's people that had a good will for him, entered into a conversation on his present undertaking and prospects: while Bradford, not discovering that he was the other printer's father, on Keimer's saying he expected soon to get the greatest part of the business into his own hands, drew him on by artful questions and starting little doubts to explain all his views, what interest he relied on, and in what manner he intended to proceed. I, who stood by and heard all, saw immediately that one of them was a crafty old sophister,[2] and the other a mere novice. Bradford left me with Keimer, who was greatly surprised when I told him who the old man was.

Keimer's printing-house, I found, consisted of an old shattered press, and one small, worn-out font [3] of English, which he was then using himself, composing an elegy on Aquila Rose, before mentioned, an ingenious young man of excellent character, much respected in the town, clerk of the Assembly, and a pretty poet. Keimer made verses too, but very indifferently. He could not be said to write them, for his manner was to compose them in the types

[1] An adjustable metal frame in which type is set.
[2] Deceiver.   [3] A set of type.

Autobiography

directly out of his head. So there being no copy, but one pair of cases,[1] and the elegy likely to require all the letter, no one could help him. I endeavored to put his press (which he had not yet used and of which he understood nothing) into order fit to be worked with; and, promising to come and print off his elegy as soon as he should have got it ready, I returned to Bradford's, who gave me a little job to do for the present, and there I lodged and dieted. A few days after Keimer sent for me to print off the elegy. And now he had got another pair of cases, and a pamphlet to reprint, on which he set me to work.

These two printers I found poorly qualified for their business. Bradford had not been bred to it and was very illiterate; and Keimer, though something of a scholar, was a mere compositor, knowing nothing of presswork. He had been one of the French prophets[2] and could act their enthusiastic agitations. At this time he did not profess any particular religion, but something of all on occasion; was very ignorant of the world, and had, as I afterward found, a good deal of the knave in his composition. He did not like my lodging at Bradford's while I worked with him. He had a house, indeed, but without furniture, so he could not lodge me; but he got me a lodging at Mr. Read's, before mentioned, who was the owner of his house; and, my chest and clothes being come by this

---
[1] Frames holding boxes of type.
[2] Protestant sect in the south of France.

time, I made rather a more respectable appearance in the eyes of Miss Read than I had done when she first happened to see me eating my roll in the street.

I began now to have some acquaintance among the young people of the town that were lovers of reading, with whom I spent my evenings very pleasantly; and gaining money by my industry and frugality, I lived very agreeably, forgetting Boston as much as I could, and not desiring that any there should know where I resided except my friend Collins, who was in my secret and kept it when I wrote to him. At length an incident happened that sent me back again much sooner than I had intended. I had a brother-in-law, Robert Holmes, master of a sloop that traded between Boston and Delaware. He, being at Newcastle forty miles below Philadelphia, heard there of me and wrote me a letter mentioning the concern of my friends in Boston at my abrupt departure, assuring me of their good will to me, and that everything would be accommodated to my mind if I would return, to which he exhorted me very earnestly. I wrote an answer to his letter, thanked him for his advice, but stated my reasons for quitting Boston fully and in such a light as to convince him I was not so wrong as he had apprehended.

Sir William Keith, Governor of the Province, was then at Newcastle, and Captain Holmes, happening to be in company with him when my letter came to hand, spoke to him of me and showed him the letter. The Governor read it and seemed surprised when he was told my age.

Autobiography

He said I appeared a young man of promising parts and therefore should be encouraged; the printers at Philadelphia were wretched ones; and, if I would set up there, he made no doubt I should succeed; for his part, he would procure me the public business, and do me every other service in his power. This my brother-in-law afterwards told me in Boston, but I knew as yet nothing of it, when one day Keimer and I being at work together near the window, we saw the Governor and another gentleman (which proved to be Colonel French of Newcastle), finely dressed, come directly across the street to our house and heard them at the door.

Keimer ran down immediately, thinking it a visit to him; but the Governor inquired for me, came up, and with a condescension and politeness I had been quite unused to, made me many compliments, desired to be acquainted with me, blamed me kindly for not having made myself known to him when I first came to the place, and would have me away with him to the tavern where he was going with Colonel French to taste, as he said, some excellent madeira. I was not a little surprised, and Keimer stared like a pig poisoned. I went, however, with the Governor and Colonel French to a tavern, at the corner of Third Street, and over the madeira he proposed my setting up my business, laid before me the probabilities of success, and both he and Colonel French assured me I should have their interest and influence in procuring the public business of both governments.[1] On my doubting whether my father would assist me in it, Sir William said he would give me a letter to him in which he would state the advantages, and he did not doubt of prevailing with him. So it was concluded I should return to Boston in the first vessel with the Governor's letter recommending me to my father. In the meantime the intention was to be kept a secret, and I went on working with Keimer as usual, the Governor sending for me now and then to dine with him, a very great honor I thought it, and conversing with me in the most affable, familiar, and friendly manner imaginable.

"The Governor Gave Me an Ample Letter . . ."

ABOUT the end of April, 1724, a little vessel offered for Boston. I took leave of Keimer as going to see my friends. The Governor gave me an ample letter, saying many flattering things of me to my father and strongly recommending the project of my setting up at Philadelphia as a thing that must make my fortune. We struck on a shoal in going down the bay and sprung a leak; we had a blustering time at sea and were obliged to pump almost continually, at which I took my turn. We arrived safe, however, at Boston in about a fortnight. I had been absent seven months, and my friends had heard nothing of me;

[1] Pennsylvania and Delaware.

for my brother Holmes was not yet returned and had not written about me. My unexpected appearance surprised the family; all were, however, very glad to see me and made me welcome, except my brother. I went to see him at his printing-house. I was better dressed than ever while in his service, having a genteel new suit from head to foot, a watch, and my pockets lined with near five pounds sterling in silver. He received me not very frankly, looked me all over, and turned to his work again.

The journeymen were inquisitive where I had been, what sort of a country it was, and how I liked it. I praised it much and the happy life I led in it, expressing strongly my intention of returning to it; and one of them asking what kind of money we had there, I produced a handful of silver and spread it before them, which was a kind of raree-show [1] they had not been used to, paper being the money of Boston. Then I took an opportunity of letting them see my watch; and lastly (my brother still grum and sullen), I gave them a piece of eight [2] to drink, and took my leave. This visit of mine offended him extremely; for when my mother some time after spoke to him of a reconciliation and of her wishes to see us on good terms together and that we might live for the future as brothers, he said I had insulted him in such a manner before his people that he could never forget or forgive it. In this, however, he was mistaken.

[1] A peep show.   [2] A Spanish dollar.

My father received the Governor's letter with some apparent surprise, but said little of it to me for some days, when, Captain Holmes returning, he showed it to him, asked him if he knew Keith and what kind of man he was, adding his opinion that he must be of small discretion to think of setting a boy up in business who wanted yet three years of being at man's estate. Holmes said what he could in favor of the project, but my father was clear in the impropriety of it and at last gave a flat denial to it. Then he wrote a civil letter to Sir William, thanking him for the patronage he had so kindly offered me, but declining to assist me as yet in setting up, I being in his opinion too young to be trusted with the management of a business so important and for which the preparation must be so expensive.

My friend and companion Collins, who was a clerk in the post office, pleased with the account I gave him of my new country, determined to go thither also; and, while I waited for my father's determination, he set out before me by land to Rhode Island, leaving his books, which were a pretty collection of mathematics and natural philosophy,[1] to come with mine and me to New York, where he proposed to wait for me.

My father, though he did not approve Sir William's proposition, was yet pleased that I had been able to obtain so advantageous a character from a person of such note where I had resided and that I had been so industrious

[1] Science.

and careful as to equip myself so handsomely in so short a time; therefore, seeing no prospect of an accommodation between my brother and me, he gave his consent to my returning again to Philadelphia, advised me to behave respectfully to the people there, endeavor to obtain the general esteem, and avoid lampooning [1] and libeling, to which he thought I had too much inclination; telling me, that by steady industry and a prudent parsimony I might save enough by the time I was one-and-twenty to set me up; and that, if I came near the matter, he would help me out with the rest. This was all I could obtain, except some small gifts as tokens of his and my mother's love, when I embarked again for New York, now with their approbation and their blessing.

The sloop putting in at Newport, Rhode Island, I visited my brother John, who had been married and settled there some years. He received me very affectionately, for he always loved me. A friend of his, one Vernon, having some money due to him in Pennsylvania, about thirty-five pounds currency, desired I would receive it for him and keep it till I had his directions what to remit it in. Accordingly, he gave me an order. This afterwards occasioned me a good deal of uneasiness.

At Newport we took in a number of passengers for New York, among which were two young women, companions, and a grave, sensible, matronlike Quaker woman, with her attendants. I had shown an obliging readiness

[1] Ridiculing.

to do her some little services, which impressed her, I suppose, with a degree of good will toward me; therefore, when she saw a daily growing familiarity between me and the two young women, which they appeared to encourage, she took me aside, and said, "Young man, I am concerned for thee, as thou hast no friend with thee and seems not to know much of the world or of the snares youth is exposed to; depend upon it, those are very bad women; I can see it in all their actions; and if thee art not upon thy guard, they will draw thee into some danger; they are strangers to thee, and I advise thee in a friendly concern for thy welfare to have no acquaintance with them." As I seemed at first not to think so ill of them as she did, she mentioned some things she had observed and heard that had escaped my notice, but now convinced me she was right. I thanked her for her kind advice, and promised to follow it. When we arrived at New York, they told me where they lived and invited me to come and see them; but I avoided it, and it was well I did; for the next day the captain missed a silver spoon and some other things that had been taken out of his cabin, and, knowing that these were a couple of strumpets, he got a warrant to search their lodgings, found the stolen goods, and had the thieves punished. So, though we had escaped a sunken rock, which we scraped upon in the passage, I thought this escape of rather more importance to me.

At New York I found my friend Collins, who had arrived there some time before me. We had been intimate from children and had read the same books together; but he had the advantage of more time for reading and studying and a wonderful genius for mathematical learning, in which he far outstripped me. While I lived in Boston, most of my hours of leisure for conversation were spent with him, and he continued a sober as well as an industrious lad, was much respected for his learning by several of the clergy and other gentlemen, and seemed to promise making a good figure in life. But during my absence he had acquired a habit of sotting with brandy; and I found by his own account and what I had heard from others that he had been drunk every day since his arrival at New York and behaved very oddly. He had gamed, too, and lost his money, so that I was obliged to discharge [1] his lodgings and defray his expenses to and at Philadelphia, which proved extremely inconvenient to me.

The then Governor of New York, Burnet (son of Bishop Burnet), hearing from the captain that a young man, one of his passengers, had a great many books, desired he would bring me to see him. I waited upon him accordingly and should have taken Collins with me but that he was not sober. The Governor treated me with great civility, showed me his library, which was a very large one, and we had a good deal of conversation about books and authors. This was the second governor who had

[1] Pay for.

done me the honor to take notice of me, which to a poor boy like me was very pleasing.

We proceeded to Philadelphia. I received on the way Vernon's money, without which we could hardly have finished our journey. Collins wished to be employed in some counting-house; but, whether they discovered his dramming by his breath or by his behavior, though he had some recommendations, he met with no success in any application and continued lodging and boarding at the same house with me and at my expense. Knowing I had that money of Vernon's, he was continually borrowing of me, still promising repayment as soon as he should be in business. At length he had got so much of it that I was distressed to think what I should do in case of being called on to remit it.

His drinking continued, about which we sometimes quarreled; for when a little intoxicated he was very fractious. Once in a boat on the Delaware with some other young men, he refused to row in his turn. "I will be rowed home," says he. "We will not row you," says I. "You must, or stay all night on the water," says he; "just as you please." The others said, "Let us row; what signifies it?" But, my mind being soured with his other conduct, I continued to refuse. So he swore he would make me row or throw me overboard; and coming along, stepping on the thwarts,[1] toward me, when he came up and struck at me, I clapped my hand under his crotch, and, rising,

[1] Seats.

pitched him head foremost into the river. I knew he was a good swimmer and so was under little concern about him; but before he could get round to lay hold of the boat, we had with a few strokes pulled her out of his reach; and ever when he drew near the boat, we asked if he would row, striking a few strokes to slide her away from him. He was ready to die with vexation and obstinately would not promise to row. However, seeing him at last beginning to tire, we lifted him in and brought him home dripping wet in the evening. We hardly exchanged a civil word afterwards, and a West India captain, who had a commission to procure a tutor for the sons of a gentleman at Barbados, happening to meet with him, agreed to carry him thither. He left me then, promising to remit me the first money he should receive in order to discharge the debt; but I never heard of him after.

"Discretion Did Not Always Accompany Years . . ."

THE BREAKING into this money of Vernon's was one of the first great errata of my life; and this affair showed that my father was not much out in his judgment when he supposed me too young to manage business of importance. But Sir William, on reading his letter, said he was too prudent. There was great difference in persons; and discretion did not always accompany years, nor was youth always without it. "And since he will not set you up," says he, "I will do it myself. Give me an inventory of the things necessary to be had from England, and I will send for them. You shall repay me when you are able; I am resolved to have a good printer here, and I am sure you must succeed." This was spoken with such an appearance of cordiality that I had not the least doubt of his meaning what he said. I had hitherto kept the proposition of my setting up a secret in Philadelphia, and I still kept it. Had it been known that I depended on the Governor, probably some friend that knew him better would have advised me not to rely on him, as I afterwards heard it as his known character to be liberal of promises which he never meant to keep. Yet unsolicited as he was by me, how could I think his generous offers insincere? I believed him one of the best men in the world.

I presented him an inventory of a little printing-house, amounting by my computation to about one hundred pounds sterling. He liked it but asked me if my being on the spot in England to choose the types and see that everything was good of the kind might not be of some advantage. "Then," says he, "when there, you may make acquaintances and establish correspondences in the bookselling and stationery way." I agreed that this might be advantageous. "Then," says he, "get yourself ready to go with Annis,"[1] which was the annual ship and the only one at that time usually passing between London and Philadelphia. But it would be some months before Annis sailed.

[1] Captain Annis commanded the ship *London-Hope*.

so I continued working with Keimer, fretting about the money Collins had got from me and in daily apprehensions of being called upon by Vernon, which, however, did not happen for some years after.

I believe I have omitted mentioning that in my first voyage from Boston, being becalmed off Block Island, our people set about catching cod, and hauled up a great many. Hitherto I had stuck to my resolution of not eating animal food, and on this occasion I considered with my master Tryon the taking every fish as a kind of unprovoked murder, since none of them had or ever could do us any injury that might justify the slaughter. All this seemed very reasonable. But I had formerly been a great lover of fish, and when this came hot out of the frying-pan, it smelt admirably well. I balanced some time between principle and inclination till I recollected that when the fish were opened, I saw smaller fish taken out of their stomachs; then thought I, "If you eat one another, I don't see why we mayn't eat you." So I dined upon cod very heartily and continued to eat with other people, returning only now and then occasionally to a vegetable diet. So convenient a thing is it to be a *reasonable creature*, since it enables one to find or make a reason for everything one has a mind to do.

Keimer and I lived on a pretty good familiar footing and agreed tolerably well, for he suspected nothing of my setting up. He retained a great deal of his old enthusiasms and loved argumentation. We therefore had many disputations. I used to work him so with my Socratic method and had trepanned [1] him so often by questions apparently so distant from any point we had in hand and yet by degrees led to the point and brought him into difficulties and contradictions that at last he grew ridiculously cautious, and would hardly answer me the most common question without asking first, "What do you intend to infer from that?" However, it gave him so high an opinion of my abilities in the confuting way that he seriously proposed my being his colleague in a project he had of setting up a new sect. He was to preach the doctrines, and I was to confound all opponents. When he came to explain with me upon the doctrines, I found several conundrums which I objected to unless I might have my way a little too and introduce some of mine.

Keimer wore his beard at full length because somewhere in the Mosaic law it is said, "Thou shalt not mar the corners of thy beard." He likewise kept the seventh day Sabbath; and these two points were essentials with him. I disliked both; but agreed to admit them upon condition of his adopting the doctrine of using no animal food. "I doubt," said he, "my constitution will not bear that." I assured him it would and that he would be the better for it. He was usually a great glutton, and I promised myself some diversion in half starving him. He agreed to try the practice if I would keep him company. I did so, and we held it for three months. We had our victuals dressed and

[1] Trapped.

brought to us regularly by a woman in the neighborhood who had from me a list of forty dishes to be prepared for us at different times in all which there was neither fish, flesh, nor fowl, and the whim suited me the better at this time from the cheapness of it, not costing us above eighteenpence sterling each per week. I have since kept several Lents most strictly, leaving the common diet for that and that for the common abruptly, without the least inconvenience, so that I think there is little in the advice of making those changes by easy gradations. I went on pleasantly, but poor Keimer suffered grievously, tired of the project, longed for the flesh-pots of Egypt, and ordered a roast pig. He invited me and two women friends to dine with him; but it being brought too soon upon table, he could not resist the temptation, and ate the whole before we came.

"My Chief Acquaintances . . ."

I HAD MADE some courtship during this time to Miss Read. I had a great respect and affection for her and had some reason to believe she had the same for me; but, as I was about to take a long voyage and we were both very young, only a little above eighteen, it was thought most prudent by her mother to prevent our going too far at present, as a marriage, if it was to take place, would be more convenient after my return when I should be, as I expected, set up in my business. Perhaps, too, she thought my expectations not so well founded as I imagined them to be.

My chief acquaintances at this time were Charles Osborne, Joseph Watson, and James Ralph, all lovers of reading. The two first were clerks to an eminent scrivener or conveyancer[1] in the town, Charles Brockden; the other was clerk to a merchant. Watson was a pious, sensible young man of great integrity; the others rather more lax in their principles of religion, particularly Ralph, who as well as Collins had been unsettled by me, for which they both made me suffer. Osborne was sensible, candid, frank, sincere and affectionate to his friends, but in literary matters too fond of criticizing. Ralph was ingenious, genteel in his manners, and extremely eloquent; I think I never knew a prettier talker. Both of them were great admirers of poetry and began to try their hands in little pieces. Many pleasant walks we four had together on Sundays into the woods near Schuylkill where we read to one another and conferred on what we read.

Ralph was inclined to pursue the study of poetry, not doubting but he might become eminent in it and make his fortune by it, alleging that the best poets must when they first began to write make as many faults as he did. Osborne dissuaded him, assured him he had no genius for poetry, and advised him to think of nothing beyond the business he was bred to; that in the mercantile way, though he had no stock, he might by his diligence and punctuality recom-

[1] One who draws deeds, searches titles, etc.

Autobiography

mend himself to employment as a factor,[1] and in time acquire wherewith to trade on his own account. I approved the amusing one's self with poetry now and then so far as to improve one's language, but no farther.

On this it was proposed that we should each of us at our next meeting produce a piece of our own composing, in order to improve by our mutual observations, criticisms, and corrections. As language and expression were what we had in view, we excluded all considerations of invention by agreeing that the task should be a version of the eighteenth Psalm, which describes the descent of a Deity. When the time of our meeting drew nigh, Ralph called on me first and let me know his piece was ready. I told him I had been busy and, having little inclination, had done nothing. He then showed me his piece for my opinion, and I much approved it as it appeared to me to have great merit. "Now," says he, "Osborne never will allow the least merit in anything of mine, but makes 1000 criticisms out of mere envy. He is not so jealous of you; I wish, therefore, you would take this piece and produce it as yours; I will pretend not to have had time and so produce nothing. We shall then see what he will say to it." It was agreed, and I immediately transcribed it that it might appear in my own hand.

We met; Watson's performance was read; there were some beauties in it, but many defects. Osborne's was read; it was much better; Ralph did it justice, remarked some

[1] Agent.

faults, but applauded the beauties. He himself had nothing to produce. I was backward, seemed desirous of being excused, had not had sufficient time to correct, etc., but no excuse could be admitted; produce I must. It was read and repeated; Watson and Osborne gave up the contest and joined in applauding it. Ralph only made some criticisms and proposed some amendments; but I defended my text. Osborne was against Ralph, and told him he was no better a critic than poet, so he dropped the argument. As they two went home together, Osborne expressed himself still more strongly in favor of what he thought my production, having restrained himself before, as he said, lest I should think it flattery. "But who would have imagined," said he, "that Franklin had been capable of such a performance; such painting, such force, such fire! He has even improved the original. In his common conversation he seems to have no choice of words; he hesitates and blunders; and yet, good God! how he writes!" When we next met, Ralph discovered [1] the trick we had played him, and Osborne was a little laughed at.

This transaction fixed Ralph in his resolution of becoming a poet. I did all I could to dissuade him from it, but he continued scribbling verses till Pope cured him.[2] He became, however, a pretty good prose writer. More of him hereafter. But, as I may not have occasion again to mention the other two, I shall just remark here that Wat-

[1] Revealed.
[2] Alexander Pope ridiculed Ralph in the *Dunciad*.

son died in my arms a few years after, much lamented, being the best of our set. Osborne went to the West Indies, where he became an eminent lawyer and made money, but died young. He and I had made a serious agreement that the one who happened first to die should if possible make a friendly visit to the other and acquaint him how he found things in that separate state. But he never fulfilled his promise.

The Governor, seeming to like my company, had me frequently to his house, and his setting me up was always mentioned as a fixed thing. I was to take with me letters recommendatory to a number of his friends, besides the letter of credit to furnish me with the necessary money for purchasing the press and types, paper, etc. For these letters I was appointed to call at different times, when they were to be ready; but a future time was still [1] named. Thus he went on till the ship, whose departure too had been several times postponed, was on the point of sailing. Then, when I called to take my leave and receive the letters, his secretary, Dr. Bard, came out to me and said the Governor was extremely busy in writing, but would be down at Newcastle before the ship, and there the letters would be delivered to me.

Ralph, though married and having one child, had determined to accompany me in this voyage. It was thought he intended to establish a correspondence and obtain goods to sell on commission; but I found afterwards that

[1] Always.

through some discontent with his wife's relations he purposed to leave her on their hands and never return again. Having taken leave of my friends and interchanged some promises with Miss Read, I left Philadelphia in the ship, which anchored at Newcastle. The Governor was there, but when I went to his lodging, the secretary came to me from him with the civilest message in the world, that he could not then see me, being engaged in business of the utmost importance, but should send the letters to me on board, wished me heartily a good voyage and a speedy return, etc. I returned on board a little puzzled, but still not doubting.

Mr. Andrew Hamilton, a famous lawyer of Philadelphia, had taken passage in the same ship for himself and son, and with Mr. Denham, a Quaker merchant, and Messrs. Onion and Russel, masters of an iron work in Maryland, had engaged the great cabin; so that Ralph and I were forced to take up with a berth in the steerage and, none on board knowing us, were considered as ordinary persons. But Mr. Hamilton and his son (it was James, since governor) returned from Newcastle to Philadelphia, the father being recalled by a great fee to plead for a seized ship; and just before we sailed, Colonel French coming on board and showing me great respect, I was more taken notice of, and with my friend Ralph invited by the other gentlemen to come into the cabin, there being now room. Accordingly, we removed thither.

Understanding that Colonel French had brought on

board the Governor's dispatches, I asked the captain for those letters that were to be under my care. He said all were put into the bag together and he could not then come at them; but before we landed in England I should have an opportunity of picking them out; so I was satisfied for the present, and we proceeded on our voyage. We had a sociable company in the cabin and lived uncommonly well, having the addition of all Mr. Hamilton's stores, who had laid in plentifully. In this passage Mr. Denham contracted a friendship for me that continued during his life. The voyage was otherwise not a pleasant one, as we had a great deal of bad weather.

When we came into the Channel, the captain kept his word with me and gave me an opportunity of examining the bag for the Governor's letters. I found none upon which my name was put as under my care. I picked out six or seven that by the handwriting I thought might be the promised letters, especially as one of them was directed to Basket, the King's printer, and another to some stationer. We arrived in London the 24th of December, 1724. I waited upon the stationer, who came first in my way, delivering the letter as from Governor Keith. "I don't know such a person," says he; but, opening the letter, "O! this is from Riddlesden. I have lately found him to be a complete rascal, and I will have nothing to do with him, nor receive any letters from him." So, putting the letter into my hand, he turned on his heel and left me to serve some customer. I was surprised to find these were not the Governor's letters; and after recollecting and comparing circumstances I began to doubt his sincerity. I found my friend Denham and opened the whole affair to him. He let me into Keith's character, told me there was not the least probability that he had written any letters for me, that no one who knew him had the smallest dependence on him; and he laughed at the notion of the Governor's giving me a letter of credit, having, as he said, no credit to give. On my expressing some concern about what I should do, he advised me to endeavor getting some employment in the way of my business. "Among the printers here," said he, "you will improve yourself, and when you return to America, you will set up to greater advantage."

We both of us happened to know as well as the stationer that Riddlesden, the attorney, was a very knave. He had half ruined Miss Read's father by persuading him to be bound [1] for him. By this letter it appeared there was a secret scheme on foot to the prejudice of Hamilton (supposed to be then coming over with us), and that Keith was concerned in it with Riddlesden. Denham, who was a friend of Hamilton's, thought he ought to be acquainted with it; so when he arrived in England, which was soon after, partly from resentment and ill will to Keith and Riddlesden and partly from good will to him, I waited on him and gave him the letter. He thanked me cordially, the information being of importance to him; and from that

[1] To sign his bond.

time he became my friend, greatly to my advantage afterwards on many occasions.

But what shall we think of a Governor's playing such pitiful tricks and imposing so grossly on a poor ignorant boy! It was a habit he had acquired. He wished to please everybody; and having little to give, he gave expectations. He was otherwise an ingenious, sensible man, a pretty good writer, and a good governor for the people, though not for his constituents, the Proprietaries, whose instructions he sometimes disregarded. Several of our best laws were of his planning and passed during his administration.

"Lodgings Together in Little Britain . . ."

RALPH and I were inseparable companions. We took lodgings together in Little Britain at three shillings and sixpence a week—as much as we could then afford. He found some relations, but they were poor and unable to assist him. He now let me know his intentions of remaining in London, and that he never meant to return to Philadelphia. He had brought no money with him, the whole he could muster having been expended in paying his passage. I had fifteen pistoles;[1] so he borrowed occasionally of me to subsist while he was looking out for business. He first endeavored to get into the playhouse, believing himself qualified for an actor; but Wilks, to whom he applied, advised him candidly not to think of that em-

[1] Spanish coins worth about $4 each.

ployment, as it was impossible he should succeed in it. Then he proposed to Roberts, a publisher in Paternoster Row, to write for him a weekly paper like the *Spectator* on certain conditions, which Roberts did not approve. Then he endeavored to get employment as a hackney writer to copy for the stationers and lawyers about the Temple, but could find no vacancy.

I immediately got into work at Palmer's, then a famous printing-house in Bartholomew Close, and here I continued near a year. I was pretty diligent, but spent with Ralph a good deal of my earnings in going to plays and other places of amusement. We had together consumed all my pistoles and now just rubbed on from hand to mouth. He seemed quite to forget his wife and child, and I, by degrees, my engagements with Miss Read, to whom I never wrote more than one letter, and that was to let her know I was not likely soon to return. This was another of the great errata of my life, which I should wish to correct if I were to live it over again. In fact, by our expenses, I was constantly kept unable to pay my passage.

At Palmer's I was employed in composing for the second edition of Wollaston's *Religion of Nature*. Some of his reasonings not appearing to me well founded, I wrote a little metaphysical[1] piece in which I made remarks on them. It was entitled *A Dissertation on Liberty and Necessity, Pleasure and Pain*. I inscribed it to my friend Ralph; I printed a small number. It occasioned my being

[1] Philosophical.

A DISSERTATION ON *Liberty* and *Necessity,* PLEASURE *and* PAIN.

*Whatever is, is in its Causes just
Since all Things are by Fate ; but purblind Man
Sees but a part o'th' Chain, the nearest Link,
His Eyes not carrying to the equal Beam
That poises all above.*

Dryd.

LONDON:
Printed in the Year MDCCXXV.

Title page of *A Dissertation on Liberty and Necessity, Pleasure and Pain,* Franklin's first book (1725). From the Huth copy in the Yale University Library. Only one other copy is known.

more considered by Mr. Palmer as a young man of some ingenuity, though he seriously expostulated with me upon the principles of my pamphlet, which to him appeared abominable.[1] My printing this pamphlet was another erratum. While I lodged in Little Britain, I made an acquaintance with one Wilcox, a bookseller, whose shop was at the next door. He had an immense collection of secondhand books. Circulating libraries were not then in use; but we agreed that on certain reasonable terms which I have now forgotten I might take, read, and return any of his books. This I esteemed a great advantage, and I made as much use of it as I could.

My pamphlet by some means falling into the hands of one Lyons, a surgeon, author of a book entitled *The Infallibility of Human Judgement*, it occasioned an acquaintance between us. He took great notice of me, called on me often to converse on those subjects, carried me to the Horns, a pale-alehouse in —— Lane, Cheapside, and introduced me to Dr. Mandeville, author of the *Fable of the Bees*, who had a club there of which he was the soul, being a most facetious, entertaining companion. Lyons, too, introduced me to Dr. Pemberton, at Batson's coffee-house, who promised to give me an opportunity some time or other of seeing Sir Isaac Newton, of which I was extremely desirous; but this never happened.

I had brought over a few curiosities, among which the

[1] Probably because it denied Free Will and doubted the immortality of the soul.

principal was a purse made of the asbestos which purifies by fire. Sir Hans Sloane heard of it, came to see me, and invited me to his house in Bloomsbury Square, where he showed me all his curiosities and persuaded me to let him add that to the number, for which he paid me handsomely.

In our house there lodged a young woman, a milliner who, I think, had a shop in the Cloisters. She had been genteelly bred, was sensible and lively, and of most pleasing conversation. Ralph read plays to her in the evenings, they grew intimate, she took another lodging, and he followed her. They lived together some time; but, he being still out of business and her income not sufficient to maintain them with her child, he took a resolution of going from London to try for a country school which he thought himself well qualified to undertake as he wrote an excellent hand and was a master of arithmetic and accounts. This, however, he deemed a business below him, and confident of future better fortune when he should be unwilling to have it known that he once was so meanly employed, he changed his name, and did me the honor to assume mine; for I soon after had a letter from him acquainting me that he was settled in a small village (in Berkshire, I think it was) where he taught reading and writing to ten or a dozen boys, at sixpence each per week, recommending Mrs. T—— to my care, and desiring me to write to him, directing for Mr. Franklin, schoolmaster, at such a place.

He continued to write frequently, sending me large

specimens of an epic poem which he was then composing and desiring my remarks and corrections. These I gave him from time to time, but endeavored rather to discourage his proceeding. One of Young's Satires was then just published. I copied and sent him a great part of it, which set in a strong light the folly of pursuing the Muses with any hope of advancement by them. All was in vain; sheets of the poem continued to come by every post. In the meantime Mrs. T——, having on his account lost her friends and business, was often in distress and used to send for me and borrow what I could spare to help her out of them. I grew fond of her company, and being at that time under no religious restraint and presuming upon my importance to her, I attempted familiarities (another erratum) which she repulsed with a proper resentment, and acquainted him with my behavior. This made a breach between us; and when he returned again to London, he let me know he thought I had canceled all the obligations he had been under to me. So I found I was never to expect his repaying me what I lent to him or advanced for him. This, however, was not then of much consequence, as he was totally unable; and in the loss of his friendship I found myself relieved from a burden. I now began to think of getting a little money beforehand, and, expecting better work, I left Palmer's to work at Watts's,[1] near

[1] The press at which Franklin worked is now in the Patent Office at Washington.

Lincoln's Inn Fields, a still greater printing-house. Here I continued all the rest of my stay in London.

"*The 'Water-American' Was Stronger . . .*"

AT MY first admission into this printing-house I took to working at press, imagining I felt a want of the bodily exercise I had been used to in America, where presswork is mixed with composing. I drank only water; the other workmen, near fifty in number, were great guzzlers of beer. On one occasion, I carried up and down stairs a large form of types in each hand, when others carried but one in both hands. They wondered to see, from this and several instances, that the *Water-American*, as they called me, was *stronger* than themselves, who drank *strong* beer! We had an alehouse boy who attended always in the house to supply the workmen. My companion at the press drank every day a pint before breakfast, a pint at breakfast with his bread and cheese, a pint between breakfast and dinner, a pint at dinner, a pint in the afternoon about six o'clock, and another when he had done his day's work. I thought it a detestable custom; but it was necessary, he supposed, to drink *strong* beer, that he might be *strong* to labor. I endeavored to convince him that the bodily strength afforded by beer could only be in proportion to the grain or flour of the barley dissolved in the water of which it was made; that there was more flour in a pennyworth of bread; and therefore, if he would eat that with a pint of water, it would give him more strength than a quart of beer. He drank on, however, and had four or five shillings to pay out of his wages every Saturday night for that muddling liquor, an expense I was free from. And thus these poor devils keep themselves always under.

Watts after some weeks desiring to have me in the composing-room, I left the pressmen; a new *bienvenu* or sum for drink, being five shillings, was demanded of me by the compositors. I thought it an imposition, as I had paid below; the master thought so too, and forbade my paying it. I stood out two or three weeks, was accordingly considered as an excommunicate, and had so many little pieces of private mischief done me, by mixing my sorts,[1] transposing my pages, breaking my matter, etc., etc., if I were ever so little out of the room, and all ascribed to the chapel[2] ghost, which they said ever haunted those not regularly admitted, that, notwithstanding the master's protection, I found myself obliged to comply and pay the money, convinced of the folly of being on ill terms with those one is to live with continually.

I was now on a fair footing with them, and soon acquired considerable influence. I proposed some reasonable

[1] Separate pieces of type.
[2] "A printing-house is always called a chapel by the workmen, the origin of which appears to have been that printing was first carried on in England in an ancient chapel converted into a printing-house, and the title has been preserved by tradition. . . . A journeyman on entering a printing-house was accustomed to pay one or more gallons of beer for the good of the chapel. This custom was falling into disuse thirty years ago; it is very properly rejected entirely in the United States."
—W. T. FRANKLIN, 1818.

alterations in their chapel laws, and carried them against all opposition. From my example, a great part of them left their muddling breakfast of beer and bread and cheese, finding they could with me be supplied from a neighboring house with a large porringer of hot water-gruel, sprinkled with pepper, crumbed with bread, and a bit of butter in it, for the price of a pint of beer, viz., three half-pence. This was a more comfortable as well as cheaper breakfast, and kept their heads clearer. Those who continued sotting with beer all day were often, by not paying, out of credit at the alehouse, and used to make interest with me to get beer, their *light*, as they phrased it, *being out*. I watched the pay-table on Saturday night and collected what I stood engaged for them, having to pay sometimes near thirty shillings a week on their accounts. This, and my being esteemed a pretty good *riggite*, that is, a jocular verbal satirist, supported my consequence in the society. My constant attendance (I never making a St. Monday [1]) recommended me to the master; and my uncommon quickness at composing occasioned my being put upon all work of dispatch, which was generally better paid. So I went on now very agreeably.

"A Widow Lady Kept the House . . ."

MY LODGING in Little Britain being too remote, I found another in Duke Street, opposite to the Romish Chapel.

[1] A Monday holiday, the origin of "blue Monday."

It was two pair of stairs backwards, at an Italian warehouse. A widow lady kept the house; she had a daughter and a maid servant and a journeyman who attended the warehouse, but lodged abroad. After sending to inquire my character at the house where I last lodged, she agreed to take me in at the same rate, 3s. 6d. per week, cheaper, as she said, from the protection she expected in having a man lodge in the house. She was a widow, an elderly woman; had been bred a Protestant, being a clergyman's daughter, but was converted to the Catholic religion by her husband, whose memory she much revered; had lived much among people of distinction, and knew a thousand anecdotes of them as far back as the times of Charles the Second. She was lame in her knees with the gout and, therefore, seldom stirred out of her room, so sometimes wanted company; and hers was so highly amusing to me that I was sure to spend an evening with her whenever she desired it. Our supper was only half an anchovy each on a very little strip of bread and butter and half a pint of ale between us; but the entertainment was in her conversation. My always keeping good hours, and giving little trouble in the family, made her unwilling to part with me; so that, when I talked of a lodging I had heard of, nearer my business, for two shillings a week, which, intent as I now was on saving money, made some difference, she bid me not think of it, for she would abate me two shillings a week for the future; so I remained with her at one shilling and sixpence as long as I stayed in England.

Autobiography

In a garret of her house there lived a maiden lady of seventy, in the most retired manner, of whom my landlady gave me this account: that she was a Roman Catholic, had been sent abroad when young and lodged in a nunnery with an intent of becoming a nun; but, the country not agreeing with her, she returned to England, where, there being no nunnery, she had vowed to lead the life of a nun, as near as might be done in those circumstances. Accordingly, she had given all her estate to charitable uses, reserving only twelve pounds a year to live on, and out of this sum she still gave a great deal in charity, living herself on water-gruel only and using no fire but to boil it. She had lived many years in that garret, being permitted to remain there gratis by successive Catholic tenants of the house below, as they deemed it a blessing to have her there. A priest visited her to confess her every day. "I have asked her," says my landlady, "how she, as she lived, could possibly find so much employment for a confessor?" "Oh," said she, "it is impossible to avoid *vain thoughts*." I was permitted once to visit her. She was cheerful and polite and conversed pleasantly. The room was clean, but had no other furniture than a mattress, a table with a crucifix and book, a stool which she gave me to sit on, and a picture over the chimney of Saint Veronica displaying her handkerchief with the miraculous figure of Christ's bleeding face on it, which she explained to me with great seriousness. She looked pale, but was never sick; and I give it as another instance on how small an income life and health may be supported.

At Watts's printing-house I contracted an acquaintance with an ingenious young man, one Wygate, who, having wealthy relations, had been better educated than most printers, was a tolerable Latinist, spoke French, and loved reading. I taught him and a friend of his to swim at twice going into the river, and they soon became good swimmers. They introduced me to some gentlemen from the country, who went to Chelsea by water to see the College [1] and Don Saltero's curiosities.[2] On our return, at the request of the company, whose curiosity Wygate had excited, I stripped and leaped into the river, and swam from near Chelsea to Blackfriars, performing on the way many feats of activity, both upon and under water, that surprised and pleased those to whom they were novelties.

I had from a child been ever delighted with this exercise, had studied and practised all Thévenot's motions and positions,[3] added some of my own, aiming at the graceful and easy as well as the useful. All these I took this occasion of exhibiting to the company and was much flattered by their admiration; and Wygate, who was desirous of becoming a master, grew more and more attached to me on that account as well as from the similarity of our studies.

---

[1] Chelsea Hospital, a veterans' home built by Wren on the site of an old theological college.

[2] A collection of freaks and relics in James Salter's coffee-house.

[3] Melchisedech Thévenot wrote *The Art of Swimming*, translated with illustrations in 1699.

Autobiography 75

He at length proposed to me traveling all over Europe together, supporting ourselves everywhere by working at our business. I was once inclined to it; but mentioning it to my good friend Mr. Denham, with whom I often spent an hour when I had leisure, he dissuaded me from it, advising me to think only of returning to Pennsylvania, which he was now about to do.

I must record one trait of this good man's character. He had formerly been in business at Bristol, but failed in debt to a number of people, compounded,[1] and went to America. There, by a close application to business as a merchant, he acquired a plentiful fortune in a few years. Returning to England in the ship with me, he invited his old creditors to an entertainment, at which he thanked them for the easy composition[2] they had favored him with, and, when they expected nothing but the treat, every man at the first remove[3] found under his plate an order on a banker for the full amount of the unpaid remainder with interest.

He now told me he was about to return to Philadelphia and should carry over a great quantity of goods in order to open a store there. He proposed to take me over as his clerk, to keep his books, in which he would instruct me, copy his letters, and attend the store. He added that as soon as I should be acquainted with mercantile business

[1] Settled by compromise.   [2] Terms of settlement.
[3] After the first course.

he would promote me by sending me with a cargo of flour and bread, etc., to the West Indies, and procure me commissions from others which would be profitable; and, if I managed well, would establish me handsomely. The thing pleased me; for I was grown tired of London, remembered with pleasure the happy months I had spent in Pennsylvania, and wished again to see it; therefore I immediately agreed on the terms of fifty pounds a year, Pennsylvania money; less, indeed, than my present gettings as a compositor, but affording a better prospect.

I now took leave of printing, as I thought, forever, and was daily employed in my new business, going about with Mr. Denham among the tradesmen to purchase various articles, and seeing them packed up, doing errands, calling upon workmen to dispatch, etc.; and, when all was on board, I had a few days' leisure. On one of these days, I was, to my surprise, sent for by a great man I knew only by name, a Sir William Wyndham, and I waited upon him. He had heard by some means or other of my swimming from Chelsea to Blackfriars and of my teaching Wygate and another young man to swim in a few hours. He had two sons about to set out on their travels; he wished to have them first taught swimming and proposed to gratify me handsomely if I would teach them. They were not yet come to town and my stay was uncertain, so I could not undertake it; but from this incident I thought it likely that, if I were to remain in England and

Autobiography

open a swimming-school, I might get a good deal of money; and it struck me so strongly that, had the overture been sooner made me, probably I should not so soon have returned to America. After many years you and I had something of more importance to do with one of these sons of Sir William Wyndham, become Earl of Egremont, which I shall mention in its place.

Thus I spent about eighteen months in London; most part of the time I worked hard at my business and spent but little upon myself except in seeing plays and in books. My friend Ralph had kept me poor; he owed me about twenty-seven pounds which I was now never likely to receive, a great sum out of my small earnings! I loved him, notwithstanding, for he had many amiable qualities. I had by no means improved my fortune; but I had picked up some very ingenious acquaintances whose conversation was of great advantage to me, and I had read considerably.

"*In Philadelphia I Found Sundry Alterations . . .*"

WE SAILED from Gravesend on the 23rd of July, 1726. For the incidents of the voyage, I refer you to my Journal, where you will find them all minutely related. Perhaps the most important part of that journal is the plan to be found in it, which I formed at sea, for regulating my future conduct in life. It is the more remarkable as being formed when I was so young, and yet being pretty faithfully adhered to quite through to old age.

We landed in Philadelphia on the 11th of October, where I found sundry alterations. Keith was no longer governor, being superseded by Major Gordon. I met him walking the streets as a common citizen. He seemed a little ashamed at seeing me, but passed without saying anything. I should have been as much ashamed at seeing Miss Read, had not her friends, despairing with reason of my return after the receipt of my letter, persuaded her to marry another, one Rogers, a potter, which was done in my absence. With him, however, she was never happy, and soon parted from him, refusing to cohabit with him or bear his name, it being now said that he had another wife. He was a worthless fellow, though an excellent workman, which was the temptation to her friends. He got into debt, ran away in 1727 or 1728, went to the West Indies, and died there. Keimer had got a better house, a shop well supplied with stationery, plenty of new types, a number of hands, though none good, and seemed to have a great deal of business.

Mr. Denham took a store in Water Street, where we opened our goods; I attended the business diligently, studied accounts, and grew, in a little time, expert at selling. We lodged and boarded together; he counseled me as a father, having a sincere regard for me. I respected and loved him, and we might have gone on together very happy; but, in the beginning of February, 1727, when I

Autobiography

had just passed my twenty-first year, we both were taken ill. My distemper was a pleurisy, which very nearly carried me off. I suffered a good deal, gave up the point in my mind, and was rather disappointed when I found myself recovering, regretting in some degree that I must now some time or other have all that disagreeable work to do over again. I forget what his distemper was; it held him a long time and at length carried him off. He left me a small legacy in a nuncupative will [1] as a token of his kindness for me, and he left me once more to the wide world; for the store was taken into the care of his executors, and my employment under him ended.

My brother-in-law, Holmes, being now at Philadelphia, advised my return to my business; and Keimer tempted me with an offer of large wages by the year to come and take the management of his printing-house that he might better attend his stationer's shop. I had heard a bad character of him in London from his wife and her friends, and was not fond of having any more to do with him. I tried for farther employment as a merchant's clerk; but, not readily meeting with any, I closed again with Keimer. I found in his house these hands: Hugh Meredith, a Welsh Pennsylvanian thirty years of age, bred to country work; honest, sensible, had a great deal of solid observation, was something of a reader, but given to drink. Stephen Potts, a young countryman of full age, bred to the same, of uncommon natural parts, and great wit and humor, but a

[1] One given by word of mouth.

little idle. These he had agreed with at extreme low wages per week, to be raised a shilling every three months, as they would deserve by improving in their business; and the expectation of these high wages, to come on hereafter, was what he had drawn them in with. Meredith was to work at press, Potts at book-binding, which he, by agreement, was to teach them, though he knew neither one nor t'other. John ——, a wild Irishman, brought up to no business, whose service for four years Keimer had purchased from the captain of a ship;[1] he, too, was to be made a pressman. George Webb, an Oxford scholar, whose time for four years he had likewise bought, intending him for a compositor, of whom more presently; and David Harry, a country boy, whom he had taken apprentice.

I soon perceived that the intention of engaging me at wages so much higher than he had been used to give was to have these raw, cheap hands formed through me; and as soon as I had instructed them, then they being all articled [2] to him, he should be able to do without me. I went on, however, very cheerfully, put his printing-house in order, which had been in great confusion, and brought his hands by degrees to mind their business and to do it better.

It was an odd thing to find an Oxford scholar in the situation of a bought servant. He was not more than

---
[1] In return for his passage.
[2] Bound by articles of apprenticeship.

eighteen years of age, and gave me this account of himself: that he was born in Gloucester, educated at a grammar-school there, had been distinguished among the scholars for some apparent superiority in performing his part, when they exhibited plays; belonged to the Witty Club there, and had written some pieces in prose and verse, which were printed in the Gloucester newspapers; thence he was sent to Oxford, where he continued about a year, but not well satisfied, wishing of all things to see London and become a player. At length, receiving his quarterly allowance of fifteen guineas, instead of discharging his debts he walked out of town, hid his gown in a furze bush, and footed it to London, where, having no friends to advise him, he fell into bad company, soon spent his guineas, found no means of being introduced among the players, grew necessitous, pawned his clothes, and wanted bread. Walking the street very hungry, and not knowing what to do with himself, a crimp's bill [1] was put into his hand, offering immediate entertainment and encouragement to such as would bind themselves to serve in America. He went directly, signed the indentures, was put into the ship, and came over, never writing a line to acquaint his friends what was become of him. He was lively, witty, good-natured, and a pleasant companion, but idle, thoughtless, and imprudent to the last degree.

John, the Irishman, soon ran away; with the rest I began

[1] Advertisement of one whose business it was to lure men into shipping as sailors.

to live very agreeably, for they all respected me the more, as they found Keimer incapable of instructing them and that from me they learned something daily. We never worked on Saturday, that being Keimer's Sabbath, so I had two days for reading. My acquaintance with ingenious people in the town increased. Keimer himself treated me with great civility and apparent regard, and nothing now made me uneasy but my debt to Vernon, which I was yet unable to pay, being hitherto but a poor economist. He, however, kindly made no demand of it.

"Quite a Factotum . . ."

OUR PRINTING-HOUSE often wanted sorts, and there was no letter-founder in America; I had seen types cast at James's in London, but without much attention to the manner; however, I now contrived a mould, made use of the letters we had as puncheons,[1] struck the matrices[2] in lead, and thus supplied in a pretty tolerable way all deficiencies. I also engraved several things on occasion; I made the ink; I was warehouseman and everything and, in short, quite a factotum.[3]

But however serviceable I might be, I found that my services became every day of less importance as the other hands improved in the business; and when Keimer paid my second quarter's wages, he let me know that he felt

[1] Stamping tools.   [2] Moulds.   [3] Jack-of-all-trades

them too heavy and thought I should make an abatement. He grew by degrees less civil, put on more of the master, frequently found fault, was captious, and seemed ready for an outbreaking. I went on, nevertheless, with a good deal of patience, thinking that his encumbered circumstances were partly the cause. At length a trifle snapped our connections; for, a great noise happening near the courthouse, I put my head out of the window to see what was the matter. Keimer, being in the street, looked up and saw me, called out to me in a loud voice and angry tone to mind my business, adding some reproachful words that nettled me the more for their publicity, all the neighbors who were looking out on the same occasion being witnesses how I was treated. He came up immediately into the printing-house, continued the quarrel, high words passed on both sides, he gave me the quarter's warning we had stipulated, expressing a wish that he had not been obliged to so long a warning. I told him his wish was unnecessary, for I would leave him that instant; and so, taking my hat, walked out of doors, desiring Meredith, whom I saw below, to take care of some things I left, and bring them to my lodgings.

Meredith came accordingly in the evening, when we talked my affair over. He had conceived a great regard for me, and was very unwilling that I should leave the house while he remained in it. He dissuaded me from returning to my native country, which I began to think of; he reminded me that Keimer was in debt for all he possessed; that his creditors began to be uneasy; that he kept his shop miserably, sold often without profit for ready money, and often trusted without keeping accounts; that he must therefore fail, which would make a vacancy I might profit of. I objected my want of money. He then let me know that his father had a high opinion of me, and from some discourse that had passed between them he was sure would advance money to set us up, if I would enter into partnership with him. "My time," says he, "will be out with Keimer in the spring; by that time we may have our press and types in from London. I am sensible I am no workman; if you like it, your skill in the business shall be set against the stock I furnish, and we will share the profits equally."

The proposal was agreeable, and I consented; his father was in town and approved of it, the more as he saw I had great influence with his son, had prevailed on him to abstain long from dram drinking, and he hoped might break him of that wretched habit entirely when we came to be so closely connected. I gave an inventory to the father, who carried it to a merchant; the things were sent for; the secret was to be kept till they should arrive, and in the meantime I was to get work, if I could, at the other printing-house. But I found no vacancy there, and so remained idle a few days, when Keimer, on a prospect of being employed to print some paper money in New Jersey, which would require cuts and various types that I only could supply, and apprehending Bradford might engage me and get the job from him, sent me a very civil

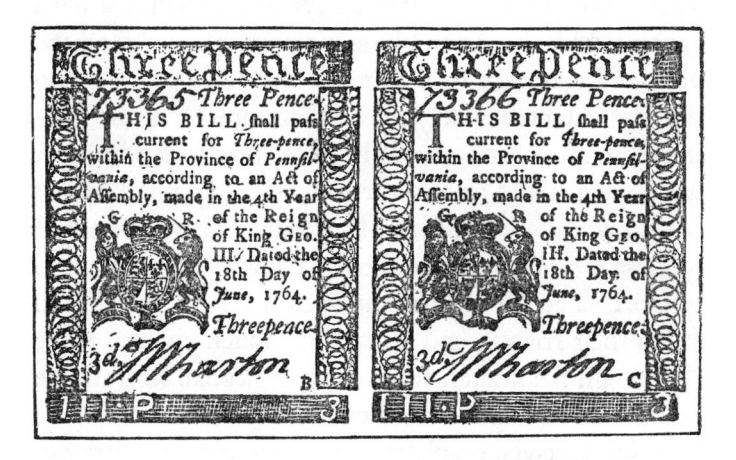

A threepence note printed by Franklin and Hall, 1764.

message that old friends should not part for a few words, the effect of sudden passion, and wishing me to return. Meredith persuaded me to comply, as it would give more opportunity for his improvement under my daily instructions; so I returned, and we went on more smoothly than for some time before. The New Jersey job was obtained, I contrived a copperplate press for it, the first that had been seen in the country; I cut several ornaments and checks for the bills. We went together to Burlington, where I executed the whole to satisfaction; and he received so large a sum for the work as to be enabled thereby to keep his head much longer above water.

At Burlington I made an acquaintance with many principal people of the Province. Several of them had been appointed by the Assembly as a committee to attend the press and take care that no more bills were printed than the law directed. They were therefore by turns constantly with us, and generally he who attended brought with him a friend or two for company. My mind having been much more improved by reading than Keimer's, I suppose it was for that reason my conversation seemed to be more valued. They had me to their houses, introduced me to their friends, and showed me much civility, while he, though the master, was a little neglected. In truth, he was an odd fish, ignorant of common life, fond of rudely opposing received opinions, slovenly to extreme dirtiness, enthusiastic [1] in some points of religion, and a little knavish withal.

We continued there near three months; and by that time I could reckon among my acquired friends, Judge Allen, Samuel Bustill, the secretary of the Province, Isaac Pearson, Joseph Cooper, and several of the Smiths, members of Assembly, and Isaac Decow, the surveyor-general. The latter was a shrewd, sagacious old man, who told me that he began for himself, when young, by wheeling clay for the brickmakers, learned to write after he was of age, carried the chain for surveyors, who taught him survey-

[1] Fanatical.

ing, and he had now by his industry acquired a good estate; and says he, "I foresee that you will soon work this man out of his business and make a fortune in it at Philadelphia." He had not then the least intimation of my intention to set up there or anywhere. These friends were afterwards of great use to me, as I occasionally was to some of them. They all continued their regard for me as long as they lived.

"With Regard to My Principles . . ."

BEFORE I enter upon my public appearance in business, it may be well to let you know the then state of my mind with regard to my principles and morals that you may see how far those influenced the future events of my life. My parents had early given me religious impressions, and brought me through my childhood piously in the Dissenting [1] way. But I was scarce fifteen, when, after doubting by turns of several points, as I found them disputed in the different books I read, I began to doubt of Revelation [2] itself. Some books against Deism fell into my hands; they were said to be the substance of sermons preached at Boyle's Lectures.[3] It happened that they wrought an effect on me quite contrary to what was intended by them; for the arguments of the Deists, which were quoted to be

[1] Dissenters were those refusing to conform to the Church of England.
[2] The Bible as a divine revelation.
[3] Robert Boyle established a lectureship "to prove the truth of the Christian religion among infidels."

refuted, appeared to me much stronger than the refutations; in short, I soon became a thorough Deist. My arguments perverted some others, particularly Collins and Ralph; but, each of them having afterwards wronged me greatly without the least compunction, and recollecting Keith's conduct towards me (who was another freethinker), and my own towards Vernon and Miss Read, which at times gave me great trouble, I began to suspect that this doctrine, though it might be true, was not very useful. My London pamphlet, which had for its motto these lines of Dryden:

> *Whatever is, is right. Though purblind man*
> *Sees but a part o' the chain, the nearest link:*
> *His eyes not carrying to the equal beam,*
> *That poises all above;*

and from the attributes of God, his infinite wisdom, goodness, and power, concluded that nothing could possibly be wrong in the world, and that vice and virtue were empty distinctions, no such things existing, appeared now not so clever a performance as I once thought it; and I doubted whether some error had not insinuated itself unperceived into my argument so as to infect all that followed, as is common in metaphysical reasonings.

I grew convinced that *truth, sincerity*, and *integrity* in dealings between man and man were of the utmost importance to the felicity of life; and I formed written resolutions, which still remain in my journal book, to

Autobiography

practise them ever while I lived. Revelation had indeed no weight with me, as such; but I entertained an opinion that, though certain actions might not be bad *because* they were forbidden by it, or good *because* it commanded them, yet probably these actions might be forbidden *because* they were bad for us, or commanded *because* they were beneficial to us in their own natures, all the circumstances of things considered. And this persuasion, with the kind hand of Providence, or some guardian angel, or accidental favorable circumstances and situations, or all together, preserved me through this dangerous time of youth and the hazardous situations I was sometimes in among strangers, remote from the eye and advice of my father, without any willful gross immorality or injustice that might have been expected from my want of religion, some foolish intrigues with low women excepted, which from the expense were rather more prejudicial to me than to them. I say willful, because the instances I have mentioned had something of *necessity* in them, from my youth, inexperience, and the knavery of others. I had therefore a tolerable character to begin the world with; I valued it properly and determined to preserve it.

We had not been long returned to Philadelphia before the new types arrived from London. We settled with Keimer and left him by his consent before he heard of it. We found a house to hire near the market and took it. To lessen the rent, which was then but twenty-four pounds a year, though I have since known it to let for seventy, we took in Thomas Godfrey, a glazier, and his family, who were to pay a considerable part of it to us, and we to board with them. We had scarce opened our letters and put our press in order, before George House, an acquaintance of mine, brought a countryman to us, whom he had met in the street inquiring for a printer. All our cash was now expended in the variety of particulars we had been obliged to procure, and this countryman's five shillings, being our first-fruits, and coming so seasonably, gave me more pleasure than any crown I have since earned; and the gratitude I felt toward House has made me often more ready than perhaps I should otherwise have been to assist young beginners.

There are croakers in every country, always boding its ruin. Such a one then lived in Philadelphia; a person of note, an elderly man, with a wise look and a very grave manner of speaking; his name was Samuel Mickle. This gentleman, a stranger to me, stopped one day at my door and asked me if I was the young man who had lately opened a new printing-house. Being answered in the affirmative, he said he was sorry for me because it was an expensive undertaking and the expense would be lost; for Philadelphia was a sinking place, the people already half-bankrupts, or near being so, all appearances to the contrary, such as new buildings and the rise of rents, being to his certain knowledge fallacious, for they were, in fact, among the things that would soon ruin us. And he gave me such a detail of misfortunes now existing, or that were soon to exist, that he left me half melancholy. Had I known him before I engaged in this business, probably I never should have done it. This man continued to live in this decaying place, and to declaim in the same strain, refusing for many years to buy a house there because all was going to destruction; and at last I had the pleasure of seeing him give five times as much for one as he might have bought it for when he first began his croaking.

"*A Club of Mutual Improvement* . . ."

I SHOULD have mentioned before that in the autumn of the preceding year I had formed most of my ingenious acquaintances into a club of mutual improvement which we called the Junto; we met on Friday evenings. The rules that I drew up required that every member in his turn should produce one or more queries on any point of morals, politics, or natural philosophy to be discussed by the company, and once in three months produce and read an essay of his own writing on any subject he pleased. Our debates were to be under the direction of a president and to be conducted in the sincere spirit of inquiry after truth, without fondness for dispute or desire of victory; and to prevent warmth all expressions of positiveness in opinions or direct contradiction were after some time made contraband and prohibited under small pecuniary penalties.

The first members were Joseph Breintnal, a copier of deeds for the scriveners, a good-natured, friendly middle-aged man, a great lover of poetry, reading all he could meet with, and writing some that was tolerable; very ingenious in many little nicknackeries, and of sensible conversation.

Thomas Godfrey, a self-taught mathematician, great in his way, and afterwards inventor of what is now called Hadley's Quadrant. But he knew little out of his way, and was not a pleasing companion; as, like most great mathematicians I have met with, he expected universal precision in everything said, or was forever denying or distinguishing upon trifles, to the disturbance of all conversation. He soon left us.

Nicholas Scull, a surveyor, afterwards surveyor-general, who loved books, and sometimes made a few verses.

William Parsons, bred a shoemaker, but, loving reading, had acquired a considerable share of mathematics, which he first studied with a view to astrology, that he afterwards laughed at. He also became surveyor-general.

William Maugridge, a joiner, a most exquisite [1] mechanic, and a solid, sensible man.

Hugh Meredith, Stephen Potts, and George Webb I have characterized before.

Robert Grace, a young gentleman of some fortune,

[1] Exact or careful.

generous, lively, and witty, a lover of punning and of his friends.

And William Coleman, then a merchant's clerk, about my age, who had the coolest, clearest head, the best heart, and the exactest morals of almost any man I ever met with. He became afterwards a merchant of great note and one of our provincial judges. Our friendship continued without interruption to his death, upwards of forty years; and the club continued almost as long and was the best school of philosophy, morality, and politics that then existed in the Province; for our queries, which were read the week preceding their discussion, put us upon reading with attention upon the several subjects that we might speak more to the purpose; and here, too, we acquired better habits of conversation, everything being studied in our rules which might prevent our disgusting each other. From hence the long continuance of the club, which I shall have frequent occasion to speak further of hereafter.

But my giving this account of it here is to show something of the interest I had, every one of these exerting themselves in recommending business to us. Breintnal particularly procured us from the Quakers the printing forty sheets of their history, the rest being to be done by Keimer; and upon this we worked exceedingly hard, for the price was low. It was a folio, pro patria size, in pica, with long primer notes. I composed of it a sheet a day and Meredith worked it off at press; it was often eleven at night, and sometimes later, before I had finished my distribution for the next day's work, for the little jobs sent in by our other friends now and then put us back. But so determined I was to continue doing a sheet a day of the folio that one night when, having imposed [1] my forms, I thought my day's work over, one of them by accident was broken and two pages reduced to pi,[2] I immediately distributed and composed it over again before I went to bed; and this industry, visible to our neighbors, began to give us character and credit; particularly, I was told, that, mention being made of the new printing-office at the merchants' Every-night club, the general opinion was that it must fail, there being already two printers in the place, Keimer and Bradford; but Dr. Baird (whom you and I saw many years after at his native place, St. Andrew's in Scotland) gave a contrary opinion: "For the industry of that Franklin," says he, "is superior to anything I ever saw of the kind; I see him still at work when I go home from club, and he is at work again before his neighbors are out of bed." This struck the rest, and we soon after had offers from one of them to supply us with stationery; but as yet we did not choose to engage in shop business.

I mention this industry the more particularly and the more freely, though it seems to be talking in my own praise, that those of my posterity who shall read it may

[1] Set up in order for printing.   [2] Disorder.

know the use of that virtue when they see its effects in my favor throughout this relation.

"*A Good Paper Would Scarcely Fail . . .*"

GEORGE WEBB, who had found a female friend that lent him wherewithal to purchase his time of Keimer, now came to offer himself as a journeyman to us. We could not then employ him; but I foolishly let him know as a secret that I soon intended to begin a newspaper and might then have work for him. My hopes of success, as I told him, were founded on this: that the then only newspaper, printed by Bradford, was a paltry thing, wretchedly managed, no way entertaining, and yet was profitable to him; I therefore thought a good paper would scarcely fail of good encouragement. I requested Webb not to mention it; but he told it to Keimer, who immediately, to be beforehand with me, published proposals for printing one himself, on which Webb was to be employed. I resented this; and to counteract them, as I could not yet begin our paper, I wrote several pieces of entertainment for Bradford's paper under the title of the "Busy Body," which Breintnal continued some months. By this means the attention of the public was fixed on that paper, and Keimer's proposals, which we burlesqued and ridiculed, were disregarded. He began his paper, however, and, after carrying it on three quarters of a year with at most only ninety subscribers, he offered it to me for a trifle; and I, having been ready some time to go on with it, took it in hand directly; and it proved in a few years extremely profitable to me.[1]

I perceive that I am apt to speak in the singular number, though our partnership still continued; the reason may be that, in fact, the whole management of the business lay upon me. Meredith was no compositor, a poor pressman, and seldom sober. My friends lamented my connection with him, but I was to make the best of it.

Our first papers made a quite different appearance from any before in the Province—a better type and better printed; but some spirited remarks of my writing on the dispute then going on between Governor Burnet and the Massachusetts Assembly struck the principal people, occasioned the paper and the manager of it to be much talked of, and in a few weeks brought them all to be our subscribers.

Their example was followed by many, and our number went on growing continually. This was one of the first good effects of my having learned a little to scribble; another was that the leading men, seeing a newspaper now in the hands of one who could also handle a pen, thought it convenient to oblige and encourage me. Bradford still printed the votes, and laws, and other public business. He had printed an address of the House to the Governor,

---

[1] The Busy-Body articles appeared in Bradford's *American Weekly Mercury* in February and March, 1729. Keimer's paper was called *The Universal Instructor in All Arts and Sciences: and Pennsylvania Gazette*. When Franklin issued his first number, October 2, 1729, he changed the title to *The Pennsylvania Gazette*.

Autobiography

in a coarse, blundering manner; we reprinted it elegantly and correctly and sent one to every member. They were sensible of the difference; it strengthened the hands of our friends in the House, and they voted us their printers for the year ensuing.

Among my friends in the House I must not forget Mr. Hamilton, before mentioned, who was then returned from England and had a seat in it. He interested himself for me strongly in that instance as he did in many others afterward, continuing his patronage till his death.

Mr. Vernon about this time put me in mind of the debt I owed him, but did not press me. I wrote him an ingenuous letter of acknowledgment, craved his forbearance a little longer, which he allowed me, and as soon as I was able, I paid the principal with interest and many thanks; so that erratum was in some degree corrected.

But now another difficulty came upon me which I had never the least reason to expect. Mr. Meredith's father, who was to have paid for our printing-house according to the expectations given me, was able to advance only one hundred pounds currency, which had been paid; and a hundred more was due to the merchant, who grew impatient and sued us all. We gave bail, but saw that if the money could not be raised in time the suit must soon come to a judgment and execution, and our hopeful prospects must, with us, be ruined, as the press and letters must be sold for payment, perhaps at half price.

In this distress two true friends, whose kindness I have

never forgotten nor ever shall forget while I can remember anything, came to me separately, unknown to each other, and without any application from me offering each of them to advance me all the money that should be necessary to enable me to take the whole business upon myself if that should be practicable; but they did not like my continuing the partnership with Meredith, who, as they said, was often seen drunk in the streets and playing at low games in alehouses, much to our discredit. These two friends were William Coleman and Robert Grace. I told them I could not propose a separation while any prospect remained of the Merediths' fulfilling their part of our agreement because I thought myself under great obligations to them for what they had done and would do if they could; but, if they finally failed in their performance and our partnership must be dissolved, I should then think myself at liberty to accept the assistance of my friends.

Thus the matter rested for some time, when I said to my partner, "Perhaps your father is dissatisfied at the part you have undertaken in this affair of ours and is unwilling to advance for you and me what he would for you alone. If that is the case, tell me, and I will resign the whole to you, and go about my business." "No," said he, "my father has really been disappointed and is really unable; and I am unwilling to distress him farther. I see this is a business I am not fit for. I was bred a farmer, and it was a folly in me to come to town and put myself at thirty years of age an apprentice to learn a new trade. Many of our

Welsh people are going to settle in North Carolina, where land is cheap. I am inclined to go with them and follow my old employment. You may find friends to assist you. If you will take the debts of the company upon you, return to my father the hundred pounds he has advanced, pay my little personal debts, and give me thirty pounds and a new saddle, I will relinquish the partnership and leave the whole in your hands." I agreed to this proposal: it was drawn up in writing, signed, and sealed immediately. I gave him what he demanded, and he went soon after to Carolina, from whence he sent me next year two long letters containing the best account that had been given of that country, the climate, the soil, husbandry, etc., for in those matters he was very judicious. I printed them in the papers, and they gave great satisfaction to the public.

"Business in My Own Name . . ."

AS SOON as he was gone, I recurred to my two friends; and because I would not give an unkind preference to either, I took half of what each had offered and I wanted of one, and half of the other, paid off the company's debts, and went on with the business in my own name, advertising that the partnership was dissolved. I think this was in or about the year 1729.

About this time there was a cry among the people for more paper money, only fifteen thousand pounds being extant in the Province, and that soon to be sunk. The wealthy inhabitants opposed any addition, being against all paper currency from an apprehension that it would depreciate, as it had done in New England, to the prejudice of all creditors. We had discussed this point in our Junto, where I was on the side of an addition, being persuaded that the first small sum struck in 1723 had done much good by increasing the trade, employment, and number of inhabitants in the Province, since I now saw all the old houses inhabited and many new ones building, whereas I remembered well that when I first walked about the streets of Philadelphia eating my roll I saw most of the houses in Walnut Street between Second and Front Streets with bills on their doors, "To Be Let," and many likewise in Chestnut Street and other streets, which made me then think the inhabitants of the city were deserting it one after another.

Our debates possessed me so fully of the subject that I wrote and printed an anonymous pamphlet on it, entitled *The Nature and Necessity of a Paper Currency*. It was well received by the common people in general; but the rich men disliked it, for it increased and strengthened the clamor for more money; and they happening to have no writers among them that were able to answer it, their opposition slackened, and the point was carried by a majority in the House. My friends there, who conceived I had been of some service, thought fit to reward me by employing me in printing the money, a very profitable job and a great help to me. This was another advantage gained by my being able to write.

The utility of this currency became by time and experience so evident as never afterwards to be much disputed, so that it grew soon to fifty-five thousand pounds and in 1739 to eighty thousand pounds, since which it arose during war to upwards of three hundred and fifty thousand pounds, trade, building, and inhabitants all the while increasing, though I now think there are limits beyond which the quantity may be hurtful.

I soon after obtained through my friend Hamilton the printing of the Newcastle paper money, another profitable job as I then thought it, small things appearing great to those in small circumstances; and these to me were really great advantages, as they were great encouragements. He procured for me also the printing of the laws and votes of that government, which continued in my hands as long as I followed the business.

I now opened a little stationer's shop. I had in it blanks [1] of all sorts, the correctest that ever appeared among us, being assisted in that by my friend Breintnal. I had also paper, parchment, chapmen's books, etc. One Whitemarsh, a compositor I had known in London, an excellent workman, now came to me and worked with me constantly and diligently; and I took an apprentice, the son of Aquila Rose.

I began now gradually to pay off the debt I was under

[1] Blank forms for legal papers.

for the printing-house. In order to secure my credit and character as a tradesman, I took care not only to be in *reality* industrious and frugal, but to avoid all appearances to the contrary. I dressed plainly; I was seen at no places of idle diversion. I never went out a fishing or shooting; a book, indeed, sometimes debauched me from my work, but that was seldom, snug, and gave no scandal; and to show that I was not above my business, I sometimes brought home the paper I purchased at the store through the streets on a wheelbarrow. Thus being esteemed an industrious, thriving young man and paying duly for what I bought, the merchants who imported stationery solicited my custom, others proposed supplying me with books, and I went on swimmingly. In the meantime, Keimer's credit and business declining daily, he was at last forced to sell his printing-house to satisfy his creditors. He went to Barbados and there lived some years in very poor circumstances.

His apprentice David Harry, whom I had instructed while I worked with him, set up in his place at Philadelphia, having bought his materials. I was at first apprehensive of a powerful rival in Harry, as his friends were very able and had a good deal of interest. I therefore proposed a partnership to him which he, fortunately for me, rejected with scorn. He was very proud, dressed like a gentleman, lived expensively, took much diversion and pleasure abroad, ran in debt, and neglected his business; upon which, all business left him, and, finding nothing to do, he followed Keimer to Barbados, taking the printing-house with him. There this apprentice employed his former master as a journeyman; they quarreled often; Harry went continually behindhand, and at length was forced to sell his types and return to his country work in Pennsylvania. The person that bought them employed Keimer to use them, but in a few years he died.

There remained now no competitor with me at Philadelphia but the old one, Bradford, who was rich and easy, did a little printing now and then by straggling hands, but was not very anxious about the business. However, as he kept the post office it was imagined he had better opportunities of obtaining news; his paper was thought a better distributer of advertisements than mine, and therefore had many more, which was a profitable thing to him, and a disadvantage to me; for, though I did indeed receive and send papers by the post, yet the public opinion was otherwise, for what I did send was by bribing the riders,[1] who took them privately, Bradford being unkind enough to forbid it, which occasioned some resentment on my part; and I thought so meanly of him for it that when I afterward came into his situation I took care never to imitate it.

"Having Turned My Thoughts to Marriage . . ."

I HAD hitherto continued to board with Godfrey, who lived in part of my house with his wife and children and

[1] Postmen on horseback

had one side of the shop for his glazier's business, though he worked little, being always absorbed in his mathematics. Mrs. Godfrey projected a match for me with a relation's daughter, took opportunities of bringing us often together, till a serious courtship on my part ensued, the girl being in herself very deserving. The old folks encouraged me by continual invitations to supper and by leaving us together, till at length it was time to explain. Mrs. Godfrey managed our little treaty. I let her know that I expected as much money with their daughter as would pay off my remaining debt for the printing-house, which I believe was not then above a hundred pounds. She brought me word they had no such sum to spare; I said they might mortgage their house in the loan-office. The answer to this, after some days, was that they did not approve the match; that on inquiry of Bradford they had been informed the printing business was not a profitable one; the types would soon be worn out and more wanted; that S. Keimer and D. Harry had failed one after the other, and I should probably soon follow them; and, therefore, I was forbidden the house and the daughter shut up.

Whether this was a real change of sentiment or only artifice on a supposition of our being too far engaged in affection to retract and therefore that we should steal a marriage, which would leave them at liberty to give or withhold what they pleased, I know not; but I suspected the latter, resented it, and went no more. Mrs. Godfrey brought me afterward some more favorable accounts of

Autobiography

their disposition and would have drawn me on again; but I declared absolutely my resolution to have nothing more to do with that family. This was resented by the Godfreys; we differed and they removed, leaving me the whole house, and I resolved to take no more inmates.

But this affair having turned my thoughts to marriage, I looked round me and made overtures of acquaintance in other places; but soon found that, the business of a printer being generally thought a poor one, I was not to expect money with a wife unless with such a one as I should not otherwise think agreeable. In the meantime that hard-to-be-governed passion of youth hurried me frequently into intrigues with low women that fell in my way, which were attended with some expense and great inconvenience, besides a continual risk to my health by a distemper which of all things I dreaded, though by great good luck I escaped it. A friendly correspondence as neighbors and old acquaintances had continued between me and Mrs. Read's family, who all had a regard for me from the time of my first lodging in their house. I was often invited there and consulted in their affairs, wherein I sometimes was of service. I pitied poor Miss Read's unfortunate situation, who was generally dejected, seldom cheerful, and avoided company. I considered my giddiness and inconstancy when in London as in a great degree the cause of her unhappiness, though the mother was good enough to think the fault more her own than mine, as she had prevented our marrying before I went thither and persuaded the other match in my absence. Our mutual affection was revived, but there were now great objections to our union. The match was indeed looked upon as invalid, a preceding wife being said to be living in England; but this could not easily be proved because of the distance; and though there was a report of his death, it was not certain. Then, though it should be true, he had left many debts which his successor might be called upon to pay. We ventured, however, over all these difficulties, and I took her to wife September 1st, 1730. None of the inconveniences happened that we had apprehended; she proved a good and faithful helpmate, assisted me much by attending the shop; we throve together, and have ever mutually endeavored to make each other happy. Thus I corrected that great erratum as well as I could.

About this time, our club meeting not at a tavern but in a little room of Mr. Grace's set apart for that purpose, a proposition was made by me that, since our books were often referred to in our disquisitions upon the queries, it might be convenient to us to have them altogether where we met that upon occasion they might be consulted; and by thus clubbing our books to a common library, we should while we liked to keep them together have each of us the advantage of using the books of all the other members, which would be nearly as beneficial as if each owned the whole. It was liked and agreed to, and we filled one end of the room with such books as we could best spare. The number was not so great as we expected; and though they had been of great use, yet some inconveniences occurring for want of due care of them, the collection after about a year was separated, and each took his books home again.

And now I set on foot my first project of a public nature, that for a subscription library. I drew up the proposals, got them put into form by our great scrivener Brockden, and by the help of my friends in the Junto procured fifty subscribers of forty shillings each to begin with and ten shillings a year for fifty years, the term our company was to continue. We afterwards obtained a charter, the company being increased to one hundred; this was the mother of all the North American subscription libraries, now so numerous. It is become a great thing itself and continually increasing. These libraries have improved the general conversation of the Americans, made the common tradesmen and farmers as intelligent as most gentlemen from other countries, and perhaps have contributed in some degree to the stand so generally made throughout the Colonies in defense of their privileges.

[*Memo. Thus far was written with the intention expressed in the beginning and therefore contains several little family anecdotes of no importance to others. What follows was written many years after in compliance with the advice contained in these letters, and accordingly intended for the public. The affairs of the Revolution occasioned the interruption.*]

Letter from Mr. Abel James with Notes of my Life (Received in Paris)

"My Dear and Honored Friend: I have often been desirous of writing to thee, but could not be reconciled to the thought that the letter might fall into the hands of the British, lest some printer or busybody should publish some part of the contents and give our friend pain and myself censure.

"Some time since there fell into my hands to my great joy about twenty-three sheets in thy own handwriting containing an account of the parentage and life of thyself directed to thy son, ending in the year 1730, with which there were notes, likewise in thy writing; a copy of which I enclose, in hopes it may be a means, if thou continued it up to a later period, that the first and latter part may be put together; and if it is not yet continued, I hope thee will not delay it. Life is uncertain, as the preacher tells us; and what will the world say if kind, humane, and benevolent Ben. Franklin should leave his friends and the world deprived of so pleasing and profitable a work, a work which would be useful and entertaining not only to a few, but to millions? The influence writings under that class have on the minds of youth is very great and has nowhere appeared to me so plain as in our public friend's journals. It almost insensibly leads the youth into the resolution of endeavoring to become as good and eminent as the journalist. Should thine, for instance, when published (and I think it could not fail of it) lead the youth to equal the industry and temperance of thy early youth, what a blessing with that class would such a work be! I know of no character living nor many of them put together who has so much in his power as thyself to promote a greater spirit of industry and early attention to business, frugality, and temperance with the American youth. Not that I think the work would have no other merit and use in the world; far from it; but the first is of such vast importance that I know nothing that can equal it."

The foregoing letter and the minutes accompanying it being shown to a friend, I received from him the following:

Letter from Mr. Benjamin Vaughan
Paris, January 31, 1783

"My Dearest Sir: When I had read over your sheets of minutes of the principal incidents of your life, recovered for you by your Quaker acquaintance, I told you I would send you a letter expressing my reasons why I thought it would be useful to complete and publish it as he desired. Various concerns have for some time past prevented this letter being written, and I do not know whether it was worth any expectation; happening to be at leisure, however, at present, I shall by writing at least interest and instruct myself; but as the terms I am inclined to use may tend to offend a person of your manners, I shall only tell you how I would address any other person who was as good and as great as yourself but less diffident. I would say to him: Sir, I solicit the history of your life from the following motives: Your history is so remarkable that if you do not give it somebody else will certainly give it; and perhaps so as nearly to do as much harm as your own management of the thing might do good. It will moreover present a table of the internal circumstances of your country which will very much tend to invite to it settlers of virtuous and manly minds. And considering the eagerness with which such information is sought by them and the extent of your reputation, I do not know of a more efficacious advertisement than your biography would give. All that has happened to you is also connected with the detail of the manners and situation of a rising people; and in this respect I do not think that the writings of Cæsar and Tacitus can be more interesting to a true judge of human nature and society. But these, sir, are small reasons in my opinion compared with the chance which your life will give for the forming of future great men; and in conjunction with your Art of Virtue (which you design to publish) of improving the features of private character and consequently of aiding all happiness both public and domestic. The two works I allude to, sir,

Autobiography

will in particular give a noble rule and example of self-education. School and other education constantly proceed upon false principles and show a clumsy apparatus pointed at a false mark; but your apparatus is simple and the mark a true one; and while parents and young persons are left destitute of other just means of estimating and becoming prepared for a reasonable course in life, your discovery that the thing is in many a man's private power will be invaluable! Influence upon the private character late in life is not only an influence late in life, but a weak influence. It is in youth that we plant our chief habits and prejudices; it is in youth that we take our party [1] as to profession, pursuits, and matrimony. In youth, therefore, the turn is given; in youth the education even of the next generation is given; in youth the private and public character is determined; and the term of life extending but from youth to age, life ought to begin well from youth, and more especially before we take our party as to our principal objects. But your biography will not merely teach self-education, but the education of a wise man; and the wisest man will receive lights and improve his progress by seeing detailed the conduct of another wise man. And why are weaker men to be deprived of such helps when we see our race has been blundering on in the dark almost without a guide in this particular from the farthest trace of time? Show then, sir, how much is to be done, both to sons and fathers; and invite all wise men to become

[1] Make our decisions.

like yourself and other men to become wise. When we see how cruel statesmen and warriors can be to the human race and how absurd distinguished men can be to their acquaintance, it will be instructive to observe the instances multiply of pacific, acquiescing manners and to find how compatible it is to be great and domestic, enviable, and yet good-humored.

"The little private incidents which you will also have to relate will have considerable use, as we want above all things rules of prudence in ordinary affairs; and it will be curious to see how you have acted in these. It will be so far a sort of key to life and explain many things that all men ought to have once explained to them to give them a chance of becoming wise by foresight. The nearest thing to having experience of one's own is to have other people's affairs brought before us in a shape that is interesting; this is sure to happen from your pen; your affairs and management will have an air of simplicity or importance that will not fail to strike; and I am convinced you have conducted them with as much originality as if you had been conducting discussions in politics or philosophy; and what more worthy of experiments and system (its importance and its errors considered) than human life?

"Some men have been virtuous blindly, others have speculated fantastically, and others have been shrewd to bad purposes; but you, sir, I am sure, will give under your hand nothing but what is at the same moment wise, practical, and good. Your account of yourself (for I suppose the

Autobiography

parallel I am drawing for Dr. Franklin will hold not only in point of character, but of private history) will show that you are ashamed of no origin, a thing the more important as you prove how little necessary all origin is to happiness, virtue, or greatness. As no end likewise happens without a means, so we shall find, sir, that even you yourself framed a plan by which you became considerable; but at the same time we may see that though the event is flattering, the means are as simple as wisdom could make them, that is, depending upon nature, virtue, thought, and habit. Another thing demonstrated will be the propriety of every man's waiting for his time for appearing upon the stage of the world. Our sensations being very much fixed to the moment, we are apt to forget that more moments are to follow the first and consequently that man should arrange his conduct so as to suit the whole of a life. Your attribution appears to have been applied to your life, and the passing moments of it have been enlivened with content and enjoyment, instead of being tormented with foolish impatience or regrets. Such a conduct is easy for those who make virtue and themselves in countenance by examples of other truly great men, of whom patience is so often the characteristic. Your Quaker correspondent, sir (for here again I will suppose the subject of my letter resembling Dr. Franklin), praised your frugality, diligence, and temperance, which he considered as a pattern for all youth; but it is singular that he should have forgotten your modesty and your disinterestedness, without which you never could have waited for your advancement or found your situation in the meantime comfortable; which is a strong lesson to show the poverty of glory and the importance of regulating our minds. If this correspondent had known the nature of your reputation as well as I do, he would have said: Your former writings and measures would secure attention to your *Biography* and *Art of Virtue;* and your *Biography* and *Art of Virtue* in return would secure attention to them. This is an advantage attendant upon a various character, and which brings all that belongs to it into greater play; and it is the more useful, as perhaps more persons are at a loss for the means of improving their minds and characters than they are for the time or the inclination to do it. But there is one concluding reflection, sir, that will show the use of your life as a mere piece of biography. This style of writing seems a little gone out of vogue, and yet it is a very useful one; and your specimen of it may be particularly serviceable, as it will make a subject of comparison with the lives of various public cut-throats and intriguers, and with absurd monastic self-tormentors or vain literary triflers. If it encourages more writings of the same kind with your own and induces more men to spend lives fit to be written, it will be worth all Plutarch's *Lives* put together. But being tired of figuring to myself a character of which every feature suits only one man in the world without giving him the praise of it, I shall end my letter, my dear Dr. Franklin, with a personal application to your proper[1] self. I am earnestly desirous, then, my dear sir, that you should let the world into the traits of your genuine character, as civil broils may otherwise tend to disguise or traduce it. Considering your great age, the caution of your character, and your peculiar style of thinking, it is not likely that any one besides yourself can be sufficiently master of the facts of your life or the intentions of your mind. Besides all this, the immense revolution of the present period will necessarily turn our attention towards the author of it, and when virtuous principles have been pretended in it, it will be highly important to show that such have really influenced; and as your own character will be the principal one to receive a scrutiny, it is proper (even for its effects upon your vast and rising country, as well as upon England and upon Europe) that it should stand respectable and eternal. For the furtherance of human happiness I have always maintained that it is necessary to prove that man is not even at present a vicious and detestable animal; and still more to prove that good management may greatly amend him; and it is for much the same reason that I am anxious to see the opinion established, that there are fair characters existing among the individuals of the race; for the moment that all men without exception shall be conceived [2] abandoned, good people will cease efforts deemed to be hopeless, and perhaps think of taking their share in the scramble of life

[1] Own.   [2] Considered.

or at least of making it comfortable principally for themselves. Take then, my dear sir, this work most speedily into hand: show yourself good as you are good; temperate as you are temperate; and above all things, prove yourself as one who from your infancy have loved justice, liberty, and concord in a way that has made it natural and consistent for you to have acted as we have seen you act in the last seventeen years of your life. Let Englishmen be made not only to respect, but even to love you. When they think well of individuals in your native country, they will go nearer to thinking well of your country; and when your countrymen see themselves well thought of by Englishmen, they will go nearer to thinking well of England. Extend your views even further; do not stop at those who speak the English tongue, but after having settled so many points in nature and politics, think of bettering the whole race of men. As I have not read any part of the life in question, but know only the character that lived it, I write somewhat at hazard. I am sure, however, that the life and the treatise I allude to (on the *Art of Virtue*) will necessarily fulfill the chief of my expectations; and still more so if you stake up the measure of suiting these performances to the several views above stated. Should they even prove unsuccessful in all that a sanguine admirer of yours hopes from them, you will at least have framed pieces to interest the human mind; and whoever gives a feeling of pleasure that is innocent to man has added so much to the fair side of a life otherwise too much darkened by anxiety and too much injured by pain. In the hope, therefore, that you will listen to the prayer addressed to you in this letter, I beg to subscribe myself, my dearest sir, etc., etc.,

"BENJ. VAUGHAN."

Citizen
1730–1756

Citizen

*Continuation of the Account of My Life,
Begun at Passy, near Paris, 1784*

It is some time since I received the above letters, but I have been too busy till now to think of complying with the request they contain. It might, too, be much better done if I were at home among my papers, which would aid my memory and help to ascertain dates; but my return being uncertain and having just now a little leisure, I will endeavor to recollect and write what I can; if I live to get home it may there be corrected and improved.

Not having any copy here of what is already written, I know not whether an account is given of the means I used to establish the Philadelphia public library, which from a small beginning is now become so considerable, though I remember to have come down to near the time of that transaction (1730). I will therefore begin here with an account of it which may be struck out if found to have been already given.

At the time I established myself in Pennsylvania there was not a good bookseller's shop in any of the colonies to the southward of Boston. In New York and Philadelphia the printers were indeed stationers; they sold only paper, etc., almanacs, ballads, and a few common schoolbooks. Those who loved reading were obliged to send for their books from England; the members of the Junto had each a few. We had left the alehouse where we first met and hired a room to hold our club in. I proposed that we should all of us bring our books to that room, where they would not only be ready to consult in our conferences, but become a common benefit, each of us being at liberty to borrow such as he wished to read at home. This was accordingly done and for some time contented us.

Finding the advantage of this little collection, I proposed to render the benefit from books more common, by commencing a public subscription library. I drew a sketch of the plan and rules that would be necessary, and got a skilful conveyancer, Mr. Charles Brockden, to put the whole in form of articles of agreement to be subscribed by which each subscriber engaged to pay a certain sum down for the first purchase of books and an annual contribution for increasing them. So few were the readers at that time in Philadelphia and the majority of us so poor that I was not able with great industry to find more than fifty persons, mostly young tradesmen, willing to pay down for this purpose forty shillings each and ten shillings per annum. On this little fund we began. The books were imported; the library was opened one day in the week for lending to the subscribers on their promissory notes to pay double the value if not duly returned. The institution soon manifested its utility, was imitated by other towns and in other provinces. The libraries were augmented by donations; reading became fashionable; and our people, having no public amusements to divert their attention from study, became better acquainted with books, and in a few years were observed by strangers to be better instructed and more intelligent than people of the same rank generally are in other countries.

When we were about to sign the above-mentioned articles, which were to be binding on us, our heirs, etc., for fifty years, Mr. Brockden, the scrivener, said to us, "You are young men, but it is scarcely probable that any of you will live to see the expiration of the term fixed in the instrument." A number of us, however, are yet living; but the instrument was after a few years rendered null by a charter that incorporated and gave perpetuity to the company.

The objections and reluctances I met with in soliciting the subscriptions made me soon feel the impropriety of presenting one's self as the proposer of any useful project that might be supposed to raise one's reputation in the smallest degree above that of one's neighbors when one has need of their assistance to accomplish that project.

I therefore put myself as much as I could out of sight and stated it as a scheme of a *number of friends* who had requested me to go about and propose it to such as they thought lovers of reading. In this way my affair went on more smoothly, and I ever after practised it on such occasions; and from my frequent successes can heartily recommend it. The present little sacrifice of your vanity will afterwards be amply repaid. If it remains a while uncertain to whom the merit belongs, some one more vain than yourself will be encouraged to claim it, and then even envy will be disposed to do you justice by plucking those assumed feathers and restoring them to their right owner.

"I Spent No Time in Taverns . . ."

THIS LIBRARY afforded me the means of improvement by constant study, for which I set apart an hour or two each day and thus repaired in some degree the loss of the learned education my father once intended for me. Reading was the only amusement I allowed myself. I spent no time in taverns, games, or frolics of any kind; and my industry in my business continued as indefatigable as it was necessary. I was indebted for my printing-house; I had a young family coming on to be educated, and I had to contend for business with two printers who were established in the place before me. My circumstances, however, grew daily easier. My original habits of frugality

Autobiography

continuing and my father having among his instructions to me when a boy frequently repeated a proverb of Solomon, "Seest thou a man diligent in his calling, he shall stand before kings, he shall not stand before mean men," I from thence considered industry as a means of obtaining wealth and distinction, which encouraged me, though I did not think that I should ever literally *stand before kings*, which, however, has since happened, for I have stood before *five*, and even had the honor of sitting down with one, the King of Denmark, to dinner.

We have an English proverb that says, "He that would thrive must ask his wife." It was lucky for me that I had one as much disposed to industry and frugality as myself. She assisted me cheerfully in my business, folding and stitching pamphlets, tending shop, purchasing old linen rags for the paper-makers, etc., etc. We kept no idle servants, our table was plain and simple, our furniture of the cheapest. For instance, my breakfast was a long time bread and milk (no tea), and I ate it out of a twopenny earthen porringer with a pewter spoon. But mark how luxury will enter families and make a progress in spite of principle: being called one morning to breakfast, I found it in a china bowl with a spoon of silver! They had been bought for me without my knowledge by my wife and had cost her the enormous sum of three-and-twenty shillings, for which she had no other excuse or apology to make but that she thought *her* husband deserved a silver spoon and china bowl as well as any of his neighbors. This was the first appearance of plate and china in our house, which afterward in a course of years as our wealth increased, augmented gradually to several hundred pounds in value.

I had been religiously educated as a Presbyterian; and though some of the dogmas of that persuasion such as the eternal decrees of God, election, reprobation,[1] etc., appeared to me unintelligible, others doubtful, and I early absented myself from the public assemblies of the sect, Sunday being my studying day, I never was without some religious principles. I never doubted, for instance, the existence of the Deity; that he made the world and governed it by his Providence; that the most acceptable service of God was the doing good to man; that our souls are immortal; and that all crime will be punished and virtue rewarded, either here or hereafter. These I esteemed the essentials of every religion; and being to be found in all the religions we had in our country, I respected them all, though with different degrees of respect, as I found them more or less mixed with other articles, which, without any tendency to inspire, promote, or confirm morality, served principally to divide us and make us unfriendly to one another. This respect to all, with an opinion that the worst had some good effects, induced me to avoid all discourse that might tend to lessen the good opinion another might have of his own religion; and as our Province increased in people and new places of worship were con-

[1] Foreordination, predestination. rejection by God's decree.

Autobiography

tinually wanted and generally erected by voluntary contribution, my mite for such purpose, whatever might be the sect, was never refused.

Though I seldom attended any public worship, I had still an opinion of its propriety and of its utility when rightly conducted, and I regularly paid my annual subscription for the support of the only Presbyterian minister or meeting we had in Philadelphia. He used to visit me sometimes as a friend and admonish me to attend his administrations, and I was now and then prevailed on to do so, once for five Sundays successively. Had he been in my opinion a good preacher perhaps I might have continued, notwithstanding the occasion I had for the Sunday's leisure in my course of study; but his discourses were chiefly either polemic [1] arguments, or explications [2] of the peculiar doctrines of our sect, and were all to me very dry, uninteresting, and unedifying, since not a single moral principle was inculcated or enforced, their aim seeming to be rather to make us Presbyterians than good citizens.

At length he took for his text that verse of the fourth chapter of Philippians, "Finally, brethren, whatsoever things are true, honest, just, pure, lovely, or of good report, if there be any virtue, or any praise, think on these things." And I imagined, in a sermon on such a text, we could not miss of having some morality. But he confined

[1] Controversial.   [2] Accounts.

himself to five points only, as meant by the apostle, viz.·
1. Keeping holy the Sabbath day. 2. Being diligent in reading the holy Scriptures. 3. Attending duly the public worship. 4. Partaking of the Sacrament. 5. Paying a due respect to God's ministers. These might be all good things, but, as they were not the kind of good things that I expected from that text, I despaired of ever meeting with them from any other, was disgusted, and attended his preaching no more. I had some years before composed a little Liturgy or form of prayer for my own private use (viz., in 1728) entitled, *Articles of Belief and Acts of Religion.* I returned to the use of this and went no more to the public assemblies. My conduct might be blameable, but I leave it without attempting further to excuse it, my present purpose being to relate facts and not to make apologies for them.

"*The Bold and Arduous Project of Arriving at Moral Perfection . . .*"

IT WAS about this time I conceived the bold and arduous project of arriving at moral perfection. I wished to live without committing any fault at any time; I would conquer all that either natural inclination, custom, or company might lead me into. As I knew, or thought I knew, what was right and wrong, I did not see why I might not always do the one and avoid the other. But I soon found I had undertaken a task of more difficulty than I had imagined. While my care was employed in guarding against one fault, I was often surprised by another; habit took the advantage of inattention; inclination was sometimes too strong for reason. I concluded, at length, that the mere speculative conviction that it was our interest to be completely virtuous was not sufficient to prevent our slipping; and that the contrary habits must be broken and good ones acquired and established before we can have any dependence on a steady, uniform rectitude of conduct. For this purpose I therefore contrived the following method.

In the various enumerations of the moral virtues I had met with in my reading I found the catalogue more or less numerous, as different writers included more or fewer ideas under the same name. Temperance, for example, was by some confined to eating and drinking, while by others it was extended to mean the moderating every other pleasure, appetite, inclination, or passion, bodily or mental, even to our avarice and ambition. I proposed to myself for the sake of clearness to use rather more names with fewer ideas annexed to each than a few names with more ideas; and I concluded under thirteen names of virtues all that at that time occurred to me as necessary or desirable and annexed to each a short precept which fully expressed the extent I gave to its meaning.

These names of virtues with their precepts were:

1. TEMPERANCE

Eat not to dullness; drink not to elevation.

2. SILENCE

Speak not but what may benefit others or yourself; avoid trifling conversation.

3. ORDER

Let all your things have their places; let each part of your business have its time.

4. RESOLUTION

Resolve to perform what you ought; perform without fail what you resolve.

5. FRUGALITY

Make no expense but to do good to others or yourself; *i.e.*, waste nothing.

6. INDUSTRY

Lose no time; be always employed in something useful; cut off all unnecessary actions.

7. SINCERITY

Use no hurtful deceit; think innocently and justly, and, if you speak, speak accordingly.

8. JUSTICE

Wrong none by doing injuries or omitting the benefits that are your duty.

9. MODERATION

Avoid extremes; forbear resenting injuries so much as you think they deserve.

10. CLEANLINESS

Tolerate no uncleanliness in body, clothes, or habitation.

11. TRANQUILLITY

Be not disturbed at trifles, or at accidents common or unavoidable.

12. CHASTITY

Rarely use venery but for health or offspring, never to dullness, weakness, or the injury of your own or another's peace or reputation.

13. HUMILITY

Imitate Jesus and Socrates.

My intention being to acquire the *habitude* of all these virtues, I judged it would be well not to distract my attention by attempting the whole at once, but to fix it on one of them at a time; and, when I should be master of that, then to proceed to another, and so on, till I should have gone through the thirteen; and, as the previous acquisition of some might facilitate the acquisition of certain others, I arranged them with that view as they stand above. Temperance first, as it tends to procure that coolness and clearness of head which is so necessary where constant vigilance was to be kept up and guard maintained against the unremitting attraction of ancient habits and the force of perpetual temptations. This being acquired and established, Silence would be more easy; and my desire being to gain knowledge at the same time that I improved in virtue, and considering that in conversation it was obtained rather by the use of the ears than of the tongue, and therefore wishing to break a habit I was getting into of prattling, punning, and joking which only made me acceptable to trifling company, I gave Silence the second place. This and the next, Order, I expected would allow me more time for attending to my project and my studies. Resolution, once become habitual, would keep me firm in my endeavors to obtain all the subsequent virtues; Frugality and Industry freeing me from my remaining debt, and producing affluence and independence, would make more easy the practice of Sincerity and Justice, etc., etc. Conceiving then that agreeably to the advice of Pythagoras in his Golden Verses daily examination would be necessary, I contrived the following method for conducting that examination.

I made a little book in which I allotted a page for each of the virtues. I ruled each page with red ink so as to have seven columns, one for each day of the week, marking each column with a letter for the day. I crossed these columns with thirteen red lines, marking the beginning of each line with the first letter of one of the virtues,

Form of the Pages

	S.	M.	T.	W.	T.	F.	S.
TEMPERANCE.							
Eat not to dullness; drink not to elevation.							
T.							
S.	*	*		*		*	
O.	* *	*	*		*	*	*
R.			*			*	
F.		*			*		
I.			*				
S.							
J.							
M.							
C.							
T.							
C.							
H.							

on which line and in its proper column I might mark by a little black spot, every fault I found upon examination to have been committed respecting that virtue upon that day.

I determined to give a week's strict attention to each of the virtues successively. Thus in the first week my great guard was to avoid every[1] the least offense against Temperance, leaving the other virtues to their ordinary chance, only marking every evening the faults of the day. Thus, if in the first week I could keep my first line, marked T, clear of spots, I supposed the habit of that virtue so much strengthened and its opposite weakened that I might venture extending my attention to include the next, and for the following week keep both lines clear of spots. Proceeding thus to the last, I could go through a course complete in thirteen weeks and four courses in a year. And like him who, having a garden to weed, does not attempt to eradicate all the bad herbs at once, which would exceed his reach and his strength, but works on one of the beds at a time, and, having accomplished the first, proceeds to a second, so I should have, I hoped, the encouraging pleasure of seeing on my pages the progress I made in virtue by clearing successively my lines of their spots till in the end by a number of courses I should be happy in viewing a clean book after a thirteen-weeks' daily examination.

This my little book had for its motto these lines from Addison's *Cato:*

[1] Even.

Autobiography

*Here will I hold. If there's a power above us
(And that there is, all nature cries aloud
Through all her works) He must delight in virtue;
And that which he delights in must be happy.*

Another from Cicero:

"*O vitæ philosophia dux! O virtutum indagatrix expultrixque vitiorum! Unus dies, bene et ex præceptis tuis actus, peccanti immortalitati est anteponendus.*" ["O Philosophy, Guide of life! O Seeker of virtue and Banisher of vice! One day well spent according to thy laws is better than an eternity of sin."]

Another from the Proverbs of Solomon, speaking of wisdom or virtue:

"Length of days is in her right hand, and in her left hand riches and honor. Her ways are ways of pleasantness, and all her paths are peace."—iii. 16, 17.

And conceiving God to be the fountain of wisdom, I thought it right and necessary to solicit his assistance for obtaining it; to this end I formed the following little prayer, which was prefixed to my tables of examination for daily use.

"O powerful Goodness! bountiful Father! merciful Guide! Increase in me that wisdom which discovers my truest interest. Strengthen my resolutions to perform what that wisdom dictates. Accept my kind offices to Thy other children as the only return in my power for Thy continual favors to me."

I used also sometimes a little prayer which I took from Thomson's poems, viz.:

Father of light and life, thou Good Supreme!
O teach me what is good; teach me Thyself!
Save me from folly, vanity, and vice,
From every low pursuit; and fill my soul
With knowledge, conscious peace, and virtue pure,
Sacred, substantial, never-fading bliss!

The precept of Order requiring that *every part of my business should have its allotted time*, one page in my little book contained the following scheme of employment for the twenty-four hours of a natural day.

THE MORNING. Question. What good shall I do this day?	5 6 7	Rise, wash, and address *Powerful Goodness!* Contrive day's business and take the resolution of the day; prosecute the present study, and breakfast.
	8 9 10 11	Work.
NOON.	12 1	Read, or overlook my accounts, and dine.
	2 3 4 5	Work.

EVENING. *Question.* What good have I done to-day?	6 7 8 9	Put things in their places. Supper. Music or diversion, or conversation. Examination of the day.
NIGHT.	10 11 12 1 2 3 4	Sleep.

I entered upon the execution of this plan for self-examination and continued it with occasional intermissions for some time. I was surprised to find myself so much fuller of faults than I had imagined; but I had the satisfaction of seeing them diminish. To avoid the trouble of renewing now and then my little book, which, by scraping out the marks on the paper of old faults to make room for new ones in a new course, became full of holes, I transferred my tables and precepts to the ivory leaves of a memorandum book on which the lines were drawn with red ink that made a durable stain, and on those lines I marked my faults with a black-lead pencil, which marks I could easily wipe out with a wet sponge. After a while I went through one course only in a year, and afterward only one in several years, till at length I omitted them entirely, being employed in voyages and business abroad with a multiplicity of affairs that interfered; but I always carried my little book with me.

"A Speckled Axe Was Best . . ."

MY SCHEME of Order gave me the most trouble; and I found that though it might be practicable where a man's business was such as to leave him the disposition of his time, that of a journeyman printer, for instance, it was not possible to be exactly observed by a master who must mix with the world and often receive people of business at their own hours. Order, too, with regard to places for things, papers, etc., I found extremely difficult to acquire. I had not been early accustomed to it, and, having an exceeding good memory, I was not so sensible of the inconvenience attending want of method. This article, therefore, cost me so much painful attention, and my faults in it vexed me so much, and I made so little progress in amendment, and had such frequent relapses, that I was almost ready to give up the attempt and content myself with a faulty character in that respect, like the man who, in buying an axe of a smith, my neighbor, desired to have the whole of its surface as bright as the edge. The smith consented to grind it bright for him if he would turn the wheel; he turned while the smith pressed the broad face of the axe hard and heavily on the stone, which made the turning of it very fatiguing. The man came every now

Autobiography

and then from the wheel to see how the work went on, and at length would take his axe as it was, without farther grinding. "No," said the smith, "turn on, turn on; we shall have it bright by-and-by; as yet, it is only speckled." "Yes," says the man, "*but I think I like a speckled axe best.*" And I believe this may have been the case with many, who, having for want of some such means as I employed found the difficulty of obtaining good and breaking bad habits in other points of vice and virtue, have given up the struggle and concluded that *a speckled axe was best;* for something that pretended to be reason was every now and then suggesting to me that such extreme nicety [1] as I exacted of myself might be a kind of foppery [2] in morals which if it were known would make me ridiculous; that a perfect character might be attended with the inconvenience of being envied and hated; and that a benevolent man should allow a few faults in himself to keep his friends in countenance.

In truth, I found myself incorrigible with respect to Order; and now I am grown old and my memory bad, I feel very sensibly the want of it. But on the whole, though I never arrived at the perfection I had been so ambitious of obtaining, but fell far short of it, yet I was by the endeavor a better and a happier man than I otherwise should have been if I had not attempted it; as those who aim at perfect writing by imitating the engraved copies, though

[1] Delicacy of discrimination. [2] Foolish vanity.

they never reach the wished-for excellence of those copies, their hand is mended by the endeavor and is tolerable while it continues fair and legible.

It may be well my posterity should be informed that to this little artifice with the blessing of God their ancestor owed the constant felicity of his life down to his 79th year, in which this is written. What reverses may attend the remainder is in the hand of Providence; but if they arrive, the reflection on past happiness enjoyed ought to help his bearing them with more resignation. To Temperance he ascribes his long-continued health and what is still left to him of a good constitution; to Industry and Frugality, the early easiness of his circumstances and acquisition of his fortune, with all that knowledge that enabled him to be a useful citizen and obtained for him some degree of reputation among the learned; to Sincerity and Justice, the confidence of his country and the honorable employs it conferred upon him; and to the joint influence of the whole mass of the virtues, even in the imperfect state he was able to acquire them, all that evenness of temper and that cheerfulness in conversation, which makes his company still sought for and agreeable even to his younger acquaintance. I hope, therefore, that some of my descendants may follow the example and reap the benefit.

It will be remarked that, though my scheme was not wholly without religion, there was in it no mark of any of the distinguishing tenets of any particular sect. I had

Autobiography

purposely avoided them; for, being fully persuaded of the utility and excellency of my method and that it might be serviceable to people in all religions, and intending some time or other to publish it, I would not have anything in it that should prejudice any one of any sect against it. I purposed writing a little comment on each virtue in which I would have shown the advantages of possessing it and the mischiefs attending its opposite vice; and I should have called my book *The Art of Virtue*, because it would have shown the means and manner of obtaining virtue, which would have distinguished it from the mere exhortation to be good that does not instruct and indicate the means, but is like the apostle's man of verbal charity, who only, without showing to the naked and hungry how or where they might get clothes or victuals, exhorted them to be fed and clothed.—*James* ii. 15, 16.

But it so happened that my intention of writing and publishing this comment was never fulfilled. I did, indeed, from time to time put down short hints of the sentiments, reasonings, etc., to be made use of in it, some of which I have still by me; but the necessary close attention to private business in the earlier part of my life and public business since have occasioned my postponing it; for, it being connected in my mind with *a great and extensive project* that required the whole man to execute and which an unforeseen succession of employs prevented my attending to, it has hitherto remained unfinished.

"I Added Humility to My List . . ."

IN THIS piece it was my design to explain and enforce this doctrine, that vicious actions are not hurtful because they are forbidden, but forbidden because they are hurtful, the nature of man alone considered; that it was, therefore, every one's interest to be virtuous who wished to be happy even in this world; and I should from this circumstance (there being always in the world a number of rich merchants, nobility, states, and princes, who have need of honest instruments for the management of their affairs, and such being so rare), have endeavored to convince young persons that no qualities were so likely to make a poor man's fortune as those of probity and integrity.

My list of virtues contained at first but twelve; but a Quaker friend having kindly informed me that I was generally thought proud; that my pride showed itself frequently in conversation; that I was not content with being in the right when discussing any point, but was overbearing, and rather insolent, of which he convinced me by mentioning several instances; I determined endeavoring to cure myself, if I could, of this vice or folly among the rest, and I added Humility to my list, giving an extensive meaning to the word.

I cannot boast of much success in acquiring the *reality* of this virtue, but I had a good deal with regard to the *appearance* of it. I made it a rule to forbear all direct contradiction to the sentiments of others and all positive assertion of my own. I even forbade myself, agreeably to the old laws of our Junto, the use of every word or expression in the language that imported [1] a fixed opinion, such as "certainly," "undoubtedly," etc., and I adopted, instead of them, "I conceive," "I apprehend," or "I imagine" a thing to be so or so; or it "so appears to me at present." When another asserted something that I thought an error, I denied myself the pleasure of contradicting him abruptly and of showing immediately some absurdity in his proposition; and in answering I began by observing that in certain cases or circumstances his opinion would be right, but in the present case there *appeared* or *seemed* to me some difference, etc. I soon found the advantage of this change in my manner; the conversations I engaged in went on more pleasantly. The modest way in which I proposed my opinions procured them a readier reception and less contradiction; I had less mortification when I was found to be in the wrong, and I more easily prevailed with others to give up their mistakes and join with me when I happened to be in the right.

And this mode, which I at first put on with some violence to natural inclination, became at length so easy, and so habitual to me that perhaps for these fifty years past no one has ever heard a dogmatical expression escape me. And to this habit (after my character of integrity)

[1] Indicated.

I think it principally owing that I had early so much weight with my fellow-citizens when I proposed new institutions or alterations in the old and so much influence in public councils when I became a member; for I was but a bad speaker, never eloquent, subject to much hesitation in my choice of words, hardly correct in language, and yet I generally carried my points.

In reality there is, perhaps, no one of our natural passions so hard to subdue as Pride. Disguise it, struggle with it, beat it down, stifle it, mortify it as much as one pleases, it is still alive, and will every now and then peep out and show itself; you will see it, perhaps, often in this history; for even if I could conceive that I had completely overcome it, I should probably be proud of my humility.

[*Thus far written at Passy, 1784.*]

"A United Party for Virtue . . ."

[*I am now about to write at home, August, 1788, but cannot have the help expected from my papers, many of them being lost in the war. I have, however, found the following.*]

HAVING mentioned *a great and extensive project* which I had conceived, it seems proper that some account should be here given of that project and its object. Its first rise in my mind appears in the following little paper, accidentally preserved, viz.:

Observations on my reading history, in Library, May 19th, 1731.

"That the great affairs of the world, the wars, revolutions, etc., are carried on and effected by parties.

"That the view of these parties is their present general interest, or what they take to be such.

"That the different views of these different parties occasion all confusion.

"That while a party is carrying on a general design, each man has his particular private interest in view.

"That as soon as a party has gained its general point, each member becomes intent upon his particular interest, which, thwarting others, breaks that party into divisions, and occasions more confusion.

"That few in public affairs act from a mere view of the good of their country, whatever they may pretend; and, though their actings bring real good to their country, yet men primarily considered that their own and their country's interest was united, and did not act from a principle of benevolence.

"That fewer still in public affairs act with a view to the good of mankind.

"There seems to me at present to be great occasion for raising a United Party for Virtue by forming the virtuous and good men of all nations into a regular body to be governed by suitable good and wise rules, which good and wise men may probably be more unanimous in their obedience to, than common people are to common laws.

"I at present think that whoever attempts this aright and is well qualified cannot fail of pleasing God and of meeting with success. B. F."

Revolving this project in my mind as to be undertaken hereafter when my circumstances should afford me the necessary leisure, I put down from time to time on pieces of paper such thoughts as occurred to me respecting it. Most of these are lost; but I find one purporting to be the substance of an intended creed, containing, as I thought, the essentials of every known religion and being free of everything that might shock the professors [1] of any religion. It is expressed in these words, viz.:

"That there is one God, who made all things.

"That He governs the world by His providence.

"That He ought to be worshiped by adoration, prayer, and thanksgiving.

"But that the most acceptable service of God is doing good to man.

"That the soul is immortal.

"And that God will certainly reward virtue and punish vice, either here or hereafter."

My ideas at that time were that the sect should be begun and spread at first among young and single men only; that each person to be initiated should not only declare his

[1] Believers.

Autobiography

assent to such creed, but should have exercised himself with the thirteen weeks' examination and practice of the virtues as in the before-mentioned model; that the existence of such a society should be kept a secret till it was become considerable, to prevent solicitations for the admission of improper persons, but that the members should each of them search among his acquaintance for ingenuous, well-disposed youths, to whom with prudent caution the scheme should be gradually communicated; that the members should engage to afford their advice, assistance, and support to each other in promoting one another's interests, business, and advancement in life; that for distinction we should be called The Society of the Free and Easy: free as being by the general practice and habit of the virtues free from the dominion of vice; and particularly by the practice of industry and frugality, free from debt, which exposes a man to confinement and a species of slavery to his creditors.

This is as much as I can now recollect of the project, except that I communicated it in part to two young men, who adopted it with some enthusiasm; but my then narrow circumstances and the necessity I was under of sticking close to my business occasioned my postponing the further prosecution of it at that time; and my multifarious occupations, public and private, induced me to continue postponing so that it has been omitted till I have no longer strength or activity left sufficient for such an enterprise; though I am still of opinion that it was a practicable scheme, and might have been very useful by forming a great number of good citizens; and I was not discouraged by the seeming magnitude of the undertaking, as I have always thought that one man of tolerable abilities may work great changes and accomplish great affairs among mankind if he first forms a good plan, and, cutting off all amusements or other employments that would divert his attention, makes the execution of that same plan his sole study and business.

"Conveying Instruction Among the Common People . . ."

IN 1732 I first published my Almanac, under the name of *Richard Saunders;* it was continued by me about twenty-five years, commonly called *Poor Richard's Almanac.* I endeavored to make it both entertaining and useful, and it accordingly came to be in such demand that I reaped considerable profit from it, vending annually near ten thousand. And observing that it was generally read, scarce any neighborhood in the Province being without it, I considered it as a proper vehicle for conveying instruction among the common people, who bought scarcely any other books; I therefore filled all the little spaces that occurred between the remarkable days in the calendar with proverbial sentences, chiefly such as inculcated industry and frugality as the means of procuring wealth and thereby securing virtue; it being more

Poor Richard, 1734.

AN Almanack

For the Year of Christ

1734,

Being the second after LEAP YEAR:

And makes since the Creation	Years
By the Account of the Eastern *Greeks*	7242
By the Latin Church, when ☉ ent. ♈	6933
By the Computation of *W.W.*	5743
By the *Roman* Chronology	5683
By the *Jewish* Rabbies	5495

Wherein is contained,

The Lunations, Eclipses, Judgment of the Weather, Spring Tides, Planets Motions & mutual Aspects, Sun and Moon's Rising and Setting; Length of Days, Time of High Water, Fairs, Courts, and observable Days.

Fitted to the Latitude of Forty Degrees, and a Meridian of Five Hours West from *London*, but may without sensible Error, serve all the adjacent Places, even from *Newfoundland* to *South-Carolina*.

By *RICHARD SAUNDERS*, Philom.

PHILADELPHIA:
Printed and sold by *B. FRANKLIN*, at the New Printing-Office near the Market.

Title page of *Poor Richard's Almanac* for 1734.

difficult for a man in want to act always honestly, as, to use here one of those proverbs, *it is hard for an empty sack to stand upright.*

These proverbs, which contained the wisdom of many ages and nations, I assembled and formed into a connected discourse prefixed to the Almanac of 1757 as the harangue of a wise old man to the people attending an auction. The bringing all these scattered counsels thus into a focus enabled them to make greater impression. The piece, being universally approved, was copied in all the newspapers of the Continent; reprinted in Britain on a broadside,[1] to be stuck up in houses; two translations were made of it in French, and great numbers bought by the clergy and gentry, to distribute gratis among their poor parishioners and tenants. In Pennsylvania, as it discouraged useless expense in foreign superfluities, some thought it had its share of influence in producing that growing plenty of money which was observable for several years after its publication.

I considered my newspaper, also, as another means of communicating instruction and in that view frequently reprinted in it extracts from the *Spectator* and other moral writers; and sometimes published little pieces of my own which had been first composed for reading in our Junto. Of these are a Socratic dialogue, tending to prove that whatever might be his parts and abilities a vicious man could not properly be called a man of sense; and a

[1] A sheet of paper printed on one side only, forming one large page.

Autobiography

discourse on self-denial, showing that virtue was not secure till its practice became a habitude and was free from the opposition of contrary inclinations. These may be found in the papers about the beginning of 1735.

In the conduct of my newspaper I carefully excluded all libeling and personal abuse, which is of late years become so disgraceful to our country. Whenever I was solicited to insert anything of that kind and the writers pleaded, as they generally did, the liberty of the press, and that a newspaper was like a stage-coach in which any one who would pay had a right to a place, my answer was that I would print the piece separately if desired and the author might have as many copies as he pleased to distribute himself, but that I would not take upon me to spread his detraction; and that, having contracted with my subscribers to furnish them with what might be either useful or entertaining, I could not fill their papers with private altercation in which they had no concern without doing them manifest injustice. Now, many of our printers make no scruple of gratifying the malice of individuals by false accusations of the fairest characters among ourselves, augmenting animosity even to the producing of duels; and are, moreover, so indiscreet as to print scurrilous reflections on the government of neighboring states and even on the conduct of our best national allies which may be attended with the most pernicious consequences. These things I mention as a caution to young printers and that they may be encouraged not to pollute their presses and disgrace their profession by such infamous practices, but refuse steadily, as they may see by my example that such a course of conduct will not, on the whole, be injurious to their interests.

In 1733 I sent one of my journeymen to Charleston, South Carolina, where a printer was wanting. I furnished him with a press and letters on an agreement of partnership by which I was to receive one-third of the profits of the business, paying one-third of the expense. He was a man of learning and honest, but ignorant in matters of account; and, though he sometimes made me remittances, I could get no account from him nor any satisfactory state [1] of our partnership while he lived. On his decease the business was continued by his widow, who, being born and bred in Holland, where, as I have been informed, the knowledge of accounts makes a part of female education, she not only sent me as clear a state as she could find of the transactions past, but continued to account with the greatest regularity and exactness every quarter afterwards, and managed the business with such success that she not only brought up reputably a family of children, but at the expiration of the term was able to purchase of me the printing-house and establish her son in it.

I mention this affair chiefly for the sake of recommending that branch of education for our young females as likely to be of more use to them and their children in case of widowhood than either music or dancing, by preserv-

[1] Statement.

Autobiography

ing them from losses by imposition of crafty men, and enabling them to continue, perhaps, a profitable mercantile house with established correspondence till a son is grown up fit to undertake and go on with it, to the lasting advantage and enriching of the family.

About the year 1734 there arrived among us from Ireland a young Presbyterian preacher named Hemphill, who delivered with a good voice, and apparently extempore, most excellent discourses which drew together considerable numbers of different persuasions who joined in admiring them. Among the rest I became one of his constant hearers, his sermons pleasing me, as they had little of the dogmatical kind, but inculcated strongly the practice of virtue or what in the religious style are called good works. Those, however, of our congregation who considered themselves as orthodox Presbyterians disapproved his doctrine and were joined by most of the old clergy, who arraigned him of heterodoxy before the synod[1] in order to have him silenced. I became his zealous partisan and contributed all I could to raise a party in his favor, and we combated for him awhile with some hopes of success. There was much scribbling pro and con upon the occasion; and finding that though an elegant preacher he was but a poor writer, I lent him my pen and wrote for him two or three pamphlets and one piece in the *Gazette* of April, 1735. Those pamphlets, as is generally the case with controversial writings, though eagerly read at the time

[1] Accused him of heresy before the clergy.

were soon out of vogue, and I question whether a single copy of them now exists.

During the contest an unlucky occurrence hurt his cause exceedingly. One of our adversaries, having heard him preach a sermon that was much admired, thought he had somewhere read the sermon before or at least a part of it. On search he found that part quoted at length in one of the British reviews from a discourse of Dr. Foster's. This detection gave many of our party disgust, who accordingly abandoned his cause, and occasioned our more speedy discomfiture in the synod. I stuck by him, however, as I rather approved his giving us good sermons composed by others than bad ones of his own manufacture, though the latter was the practice of our common teachers. He afterward acknowledged to me that none of those he preached were his own, adding that his memory was such as enabled him to retain and repeat any sermon after one reading only. On our defeat, he left us in search elsewhere of better fortune, and I quitted the congregation, never joining it after, though I continued many years my subscription for the support of its ministers.

I had begun in 1733 to study languages; I soon made myself so much a master of the French as to be able to read the books with ease. I then undertook the Italian. An acquaintance, who was also learning it, used often to tempt me to play chess with him. Finding this took up too much of the time I had to spare for study, I at length refused to play any more unless on this condition, that the victor in every game should have a right to impose a task either in parts of the grammar to be got by heart or in translations, etc., which tasks the vanquished was to perform upon honor, before our next meeting. As we played pretty equally we thus beat one another into that language. I afterwards with a little painstaking acquired as much of the Spanish as to read their books also.

I have already mentioned that I had only one year's instruction in a Latin school and that when very young, after which I neglected that language entirely. But when I had attained an acquaintance with the French, Italian, and Spanish, I was surprised to find on looking over a Latin Testament that I understood so much more of that language than I had imagined, which encouraged me to apply myself again to the study of it, and I met with more success, as those preceding languages had greatly smoothed my way.

From these circumstances, I have thought that there is some inconsistency in our common mode of teaching languages. We are told that it is proper to begin first with the Latin and, having acquired that, it will be more easy to attain those modern languages which are derived from it; and yet we do not begin with the Greek in order more easily to acquire the Latin. It is true that if you can clamber and get to the top of a staircase without using the steps, you will more easily gain them in descending; but certainly if you begin with the lowest you will with more ease ascend to the top; and I would therefore offer it to the consideration of those who superintend the education of our youth whether, since many of those who begin with the Latin quit the same after spending some years without having made any great proficiency and what they have learnt becomes almost useless so that their time has been lost, it would not have been better to have begun with the French, proceeding to the Italian, etc.; for, though after spending the same time they should quit the study of languages and never arrive at the Latin, they would, however, have acquired another tongue or two that, being in modern use, might be serviceable to them in common life.

After ten years' absence from Boston and having become easy in my circumstances, I made a journey thither to visit my relations, which I could not sooner well afford. In returning I called at Newport to see my brother, then settled there with his printing-house. Our former differences were forgotten, and our meeting was very cordial and affectionate. He was fast declining in his health, and requested of me that in case of his death, which he apprehended not far distant, I would take home his son, then but ten years of age, and bring him up to the printing business. This I accordingly performed, sending him a few years to school before I took him into the office. His mother carried on the business till he was grown up, when I assisted him with an assortment of new types, those of his father being in a manner worn out. Thus it was that I made my brother ample amends for the service I had deprived him of by leaving him so early.

In 1736 I lost one of my sons, a fine boy of four years old, by the small-pox, taken in the common way. I long regretted bitterly and still regret that I had not given it to him by inoculation. This I mention for the sake of parents who omit that operation on the supposition that they should never forgive themselves if a child died under it, my example showing that the regret may be the same either way and that, therefore, the safer should be chosen.

"The Increase of Our Influence in Public Affairs . . ."

OUR CLUB, the Junto, was found so useful and afforded such satisfaction to the members that several were desirous of introducing their friends, which could not well be done without exceeding what we had settled as a convenient number, viz., twelve. We had from the beginning made it a rule to keep our institution a secret, which was pretty well observed; the intention was to avoid applications of improper persons for admittance, some of whom perhaps we might find it difficult to refuse. I was one of those who were against any addition to our number, but instead of it made in writing a proposal that every member separately should endeavor to form a subordinate club with the same rules respecting queries, etc. and without informing them of the connection with the Junto. The advantages proposed were, the improvement of so many more young citizens by the use of our institutions; our better acquaintance with the general sentiments of the inhabitants on any occasion, as the Junto member might propose what queries we should desire and was to report to the Junto what passed in his separate club; the promotion of our particular interests in business by more extensive recommendation, and the increase of our influence in public affairs, and our power of doing good by spreading through the several clubs the sentiments of the Junto.

The project was approved, and every member undertook to form his club, but they did not all succeed. Five or six only were completed, which were called by different names as the Vine, the Union, the Band, etc. They were useful to themselves and afforded us a good deal of amusement, information, and instruction, besides answering, in some considerable degree, our views of influencing the public opinion on particular occasions, of which I shall give some instances in course of time as they happened.

My first promotion was my being chosen in 1736 clerk of the General Assembly. The choice was made that year without opposition; but the year following, when I was again proposed (the choice, like that of the members, being annual), a new member made a long speech against me in order to favor some other candidate. I was, however, chosen, which was the more agreeable to me as besides the pay for the immediate service as clerk the place gave me a better opportunity of keeping up an

Autobiography

interest among the members, which secured to me the business of printing the votes, laws, paper money, and other occasional jobs for the public that on the whole were very profitable.

I therefore did not like the opposition of this new member, who was a gentleman of fortune and education with talents that were likely to give him in time great influence in the House, which, indeed, afterwards happened. I did not, however, aim at gaining his favor by paying any servile respect to him, but after some time took this other method. Having heard that he had in his library a certain very scarce and curious book, I wrote a note to him, expressing my desire of perusing that book and requesting he would do me the favor of lending it to me for a few days. He sent it immediately, and I returned it in about a week with another note expressing strongly my sense of the favor. When we next met in the House, he spoke to me (which he had never done before) and with great civility; and he ever after manifested a readiness to serve me on all occasions, so that we became great friends and our friendship continued to his death. This is another instance of the truth of an old maxim I had learned, which says, "He that has once done you a kindness will be more ready to do you another than he whom you yourself have obliged." And it shows how much more profitable it is prudently to remove, than to resent, return, and continue inimical proceedings.

In 1737, Colonel Spotswood, late **Governor of Virginia**,

and then postmaster-general, being dissatisfied with the conduct of his deputy at Philadelphia respecting some negligence in rendering and inexactitude of his accounts, took from him the commission and offered it to me. I accepted it readily and found it of great advantage; for though the salary was small it facilitated the correspondence that improved my newspaper, increased the number demanded as well as the advertisements to be inserted, so that it came to afford me a considerable income. My old competitor's newspaper declined proportionately, and I was satisfied without retaliating his refusal while postmaster to permit my papers being carried by the riders. Thus he suffered greatly from his neglect in due accounting; and I mention it as a lesson to those young men who may be employed in managing affairs for others that they should always render accounts and make remittances with great clearness and punctuality. The character of observing such a conduct is the most powerful of all recommendations to new employments and increase of business.

"A More Effectual Watch . . ."

I BEGAN now to turn my thoughts a little to public affairs, beginning, however, with small matters. The city watch [1] was one of the first things that I conceived to want regulation. It was managed by the constables of the respective wards in turn; the constable warned a number of

[1] Police force.

housekeepers [1] to attend him for the night. Those who chose never to attend paid him six shillings a year to be excused, which was supposed to be for hiring substitutes, but was, in reality, much more than was necessary for that purpose and made the constableship a place of profit; and the constable for a little drink often got such ragamuffins about him as a watch that respectable housekeepers did not choose to mix with. Walking the rounds, too, was often neglected, and most of the nights spent in tippling. I thereupon wrote a paper to be read in Junto representing these irregularities, but insisting more particularly on the inequality of this six-shilling tax of the constables respecting the circumstances of those who paid it, since a poor widow housekeeper, all whose property to be guarded by the watch did not perhaps exceed the value of fifty pounds, paid as much as the wealthiest merchant, who had thousands of pounds' worth of goods in his stores.

On the whole I proposed as a more effectual watch the hiring of proper men to serve constantly in that business; and as a more equitable way of supporting the charge, the levying a tax that should be proportioned to the property. This idea, being approved by the Junto, was communicated to the other clubs, but as arising in each of them; and though the plan was not immediately carried into execution, yet by preparing the minds of people for the change it paved the way for the law obtained a few

[1] House owners.

years after, when the members of our clubs were grown into more influence.

About this time I wrote a paper (first to be read in Junto, but it was afterward published) on the different accidents and carelessnesses by which houses were set on fire with cautions against them and means proposed of avoiding them. This was much spoken of as a useful piece and gave rise to a project which soon followed it of forming a company for the more ready extinguishing of fires and mutual assistance in removing and securing of goods when in danger. Associates in this scheme were presently found, amounting to thirty. Our articles of agreement obliged every member to keep always in good order and fit for use a certain number of leather buckets with strong bags and baskets (for packing and transporting of goods) which were to be brought to every fire; and we agreed to meet once a month and spend a social evening together in discoursing and communicating such ideas as occurred to us upon the subjects of fires as might be useful in our conduct on such occasions.

The utility of this institution soon appeared, and many more desiring to be admitted than we thought convenient for one company, they were advised to form another, which was accordingly done; and this went on, one new company being formed after another, till they became so numerous as to include most of the inhabitants who were men of property; and now at the time of my writing this, though upward of fifty years since its establishment, that which I first formed, called the Union Fire Company, still subsists and flourishes, though the first members are all deceased but myself and one who is older by a year than I am. The small fines that have been paid by members for absence at the monthly meetings have been applied to the purchase of fire-engines, ladders, fire-hooks, and other useful implements for each company, so that I question whether there is a city in the world better provided with the means of putting a stop to beginning conflagrations; and in fact since these institutions, the city has never lost by fire more than one or two houses at a time, and the flames have often been extinguished before the house in which they began has been half consumed.

"The Reverend Mr. Whitefield . . ."

IN 1739 arrived among us from Ireland the Reverend Mr. Whitefield,[1] who had made himself remarkable there as an itinerant preacher. He was at first permitted to preach in some of our churches; but the clergy, taking a dislike to him, soon refused him their pulpits, and he was obliged to preach in the fields. The multitudes of all sects and denominations that attended his sermons were enormous, and it was matter of speculation to me, who was one of the number, to observe the extraordinary influence of his oratory on his hearers and how much they admired and respected him, notwithstanding his common abuse

[1] George Whitefield (pronounced Whit-field), Methodist evangelist.

of them by assuring them they were naturally *half beasts and half devils*. It was wonderful to see the change soon made in the manners of our inhabitants. From being thoughtless or indifferent about religion, it seemed as if all the world were growing religious, so that one could not walk through the town in an evening without hearing psalms sung in different families of every street.

And it being found inconvenient to assemble in the open air, subject to its inclemencies, the building of a house to meet in was no sooner proposed and persons appointed to receive contributions, but sufficient sums were soon received to procure the ground and erect the building, which was one hundred feet long and seventy broad, about the size of Westminster Hall; and the work was carried on with such spirit as to be finished in a much shorter time than could have been expected. Both house and ground were vested in trustees expressly for the use of any preacher of any religious persuasion who might desire to say something to the people at Philadelphia; the design in building not being to accommodate any particular sect, but the inhabitants in general; so that even if the Mufti [1] of Constantinople were to send a missionary to preach Mohammedanism to us, he would find a pulpit at his service.

Mr. Whitefield in leaving us went preaching all the way through the Colonies to Georgia. The settlement of that Province had lately been begun, but instead of being made

[1] Head of the Mohammedan religion.

with hardy, industrious husbandmen, accustomed to labor, the only people fit for such an enterprise, it was with families of broken shopkeepers and other insolvent debtors, many of indolent and idle habits, taken out of the jails, who, being set down in the woods, unqualified for clearing land and unable to endure the hardships of a new settlement, perished in numbers, leaving many helpless children unprovided for. The sight of their miserable situation inspired the benevolent heart of Mr. Whitefield with the idea of building an Orphan House there in which they might be supported and educated. Returning northward, he preached up this charity and made large collections, for his eloquence had a wonderful power over the hearts and purses of his hearers of which I myself was an instance.

I did not disapprove of the design, but as Georgia was then destitute of materials and workmen and it was proposed to send them from Philadelphia at a great expense, I thought it would have been better to have built the house here and brought the children to it. This I advised; but he was resolute in his first project, rejected my counsel, and I therefore refused to contribute. I happened soon after to attend one of his sermons in the course of which I perceived he intended to finish with a collection, and I silently resolved he should get nothing from me. I had in my pocket a handful of copper money, three or four silver dollars, and five pistoles in gold. As he proceeded I began to soften and concluded to give the coppers. Another stroke of his oratory made me ashamed of that and determined me to give the silver; and he finished so admirably that I emptied my pocket wholly into the collector's dish, gold and all. At this sermon there was also one of our club who, being of my sentiments respecting the building in Georgia and suspecting a collection might be intended, had by precaution emptied his pockets before he came from home. Towards the conclusion of the discourse, however, he felt a strong desire to give and applied to a neighbor who stood near him to borrow some money for the purpose. The application was unfortunately made to perhaps the only man in the company who had the firmness not to be affected by the preacher. His answer was, "At any other time, Friend Hopkinson, I would lend to thee freely; but not now, for thee seems to be out of thy right senses."

Some of Mr. Whitefield's enemies affected to suppose that he would apply these collections to his own private emolument; but I, who was intimately acquainted with him (being employed in printing his Sermons and Journals, etc.), never had the least suspicion of his integrity, but am to this day decidedly of opinion that he was in all his conduct a perfectly *honest man;* and methinks my testimony in his favor ought to have the more weight as we had no religious connection. He used indeed sometimes to pray for my conversion, but never had the satisfaction of believing that his prayers were heard. Ours was a mere civil friendship, sincere on both sides, and lasted to his death.

The following instance will show something of the terms on which we stood. Upon one of his arrivals from England at Boston he wrote to me that he should come soon to Philadelphia, but knew not where he could lodge when there, as he understood his old friend and host Mr. Benezet was removed to Germantown. My answer was, "You know my house; if you can make shift with its scanty accommodations, you will be most heartily welcome." He replied that if I made that kind offer for Christ's sake I should not miss of a reward. And I returned, "Don't let me be mistaken; it was not for Christ's sake, but for your sake." One of our common acquaintance jocosely remarked that, knowing it to be the custom of the saints, when they received any favor, to shift the burden of the obligation from off their own shoulders and place it in heaven, I had contrived to fix it on earth.

The last time I saw Mr. Whitefield was in London, when he consulted me about his Orphan Home concern, and his purpose of appropriating it to the establishment of a college.

He had a loud and clear voice and articulated his words and sentences so perfectly that he might be heard and understood at a great distance, especially as his auditories, however numerous, observed the most exact silence. He preached one evening from the top of the courthouse steps, which are in the middle of Market Street and on the west side of Second Street, which crosses it at right angles. Both streets were filled with his hearers to a considerable distance. Being among the hindmost in Market Street, I had the curiosity to learn how far he could be heard, by retiring backwards down the street towards the river; and I found his voice distinct till I came near Front Street, when some noise in that street obscured it. Imagining then a semicircle of which my distance should be the radius and that it were filled with auditors, to each of whom I allowed two square feet, I computed that he might well be heard by more than thirty thousand. This reconciled me to the newspaper accounts of his having preached to twenty-five thousand people in the fields and to the ancient histories of generals haranguing whole armies, of which I had sometimes doubted.

By hearing him often I came to distinguish easily between sermons newly composed and those which he had often preached in the course of his travels. His delivery of the latter was so improved by frequent repetitions that every accent, every emphasis, every modulation of voice was so perfectly well turned and well placed that without being interested in the subject one could not help being pleased with the discourse, a pleasure of much the same kind with that received from an excellent piece of music. This is an advantage itinerant preachers have over those who are stationary, as the latter cannot well improve their delivery of a sermon by so many rehearsals.

His writing and printing from time to time gave great

advantage to his enemies; unguarded expressions and even erroneous opinions delivered in preaching might have been afterwards explained or qualified by supposing others that might have accompanied them or they might have been denied; but *litera scripta manet*.[1] Critics attacked his writings violently and with so much appearance of reason as to diminish the number of his votaries [2] and prevent their increase, so that I am of opinion if he had never written anything he would have left behind him a much more numerous and important sect, and his reputation might in that case have been still growing even after his death, as, there being nothing of his writing on which to found a censure and give him a lower character, his proselytes would be left at liberty to feign for him as great a variety of excellences as their enthusiastic admiration might wish him to have possessed.

My business was now continually augmenting and my circumstances growing daily easier, my newspaper having become very profitable as being for a time almost the only one in this and the neighboring provinces. I experienced too the truth of the observation, "that after getting the first hundred pound, it is more easy to get the second," money itself being of a prolific nature.

The partnership at Carolina having succeeded, I was encouraged to engage in others and to promote several of my workmen who had behaved well by establishing them

[1] The written *letter* remains.   [2] Followers.

with printing-houses in different colonies, on the same terms with that in Carolina. Most of them did well, being enabled at the end of our term, six years, to purchase the types of me and go on working for themselves, by which means several families were raised. Partnerships often finish in quarrels; but I was happy in this, that mine were all carried on and ended amicably, owing, I think, a good deal to the precaution of having very explicitly settled in our articles everything to be done by or expected from each partner so that there was nothing to dispute, which precaution I would therefore recommend to all who enter into partnerships; for, whatever esteem partners may have for, and confidence in, each other at the time of the contract, little jealousies and disgusts may arise with ideas of inequality in the care and burden of the business, etc. which are attended often with breach of friendship and of the connection, perhaps with lawsuits and other disagreeable consequences.

"The Necessity of Union and Discipline for Our Defense . . ."

I HAD on the whole abundant reason to be satisfied with my being established in Pennsylvania. There were, however, two things that I regretted, there being no provision for defense nor for a complete education of youth; no militia nor any college. I therefore in 1743 drew up a proposal for establishing an academy; and at that time,

Autobiography

thinking the Reverend Mr. Peters, who was out of employ, a fit person to superintend such an institution, I communicated the project to him; but he, having more profitable views [1] in the service of the Proprietaries,[2] which succeeded, declined the undertaking; and not knowing another at that time suitable for such a trust, I let the scheme lie awhile dormant. I succeeded better the next year, 1744, in proposing and establishing a Philosophical Society.[3] The paper I wrote for that purpose will be found among my writings, when collected.

With respect to defense, Spain having been several years at war against Great Britain and being at length joined by France,[4] which brought us into great danger; and the labored and long-continued endeavor of our Governor, Thomas, to prevail with our Quaker Assembly to pass a militia law and make other provisions for the security of the province having proved abortive, I determined to try what might be done by a voluntary association of the people. To promote this, I first wrote and published a pamphlet entitled *Plain Truth* [5] in which I stated our defenseless situation in strong lights, with the necessity

[1] Prospects.
[2] The Colony of Pennsylvania was privately owned by the sons of William Penn, who are referred to as "the Proprietaries."
[3] Five of the ten Philadelphia members of the Society are known to have belonged to the Junto.
[4] England had been at war with Spain since 1739 over British smuggling in Spanish America and since 1744 with France over the Austrian succession.
[5] Franklin's *Plain Truth* was published November 17, 1747.

PLAIN TRUTH:

O R,

SERIOUS CONSIDERATIONS

On the PRESENT STATE of the

CITY of *PHILADELPHIA,*

A N D

PROVINCE of *PENNSYLVANIA.*

By a TRADESMAN of *Philadelphia.*

Capta urbe, nihil fit reliqui victis. Sed, per Deos immortales, vos ego appello, qui semper domos, villas, signa, tabulas vestras, tantæ æstimationis fecistis; si ista, cujuscumque modi sint, quæ amplexamini, retinere, si voluptatibus vestris otium præbere vultis; expergiscimini aliquando, & capessite rempublicam. Non agitur nunc de sociorum injuriis; LIBERTAS *&* ANIMA *nostra in dubio est. Dux hostium cum exercitu supra caput est. Vos cunctamini etiam nunc, & dubitatis quid faciatis? Scilicet, res ipsa aspera est, sed vos non timetis eam. Imo vero maxume; sed inertia & mollitia animi, alius alium exspectantes, cunctamini; videlicet, Diis immortalibus confisi, qui hanc rempublicam in maxumis periculis servavere.* NON VOTIS, NEQUE SUPPLICIIS MULIEBRIBUS, AUXILIA DEORUM PARANTUR: *vigilando, agendo, bene consulendo, prospere omnia cedunt. Ubi socordiæ tete atque ignaviæ tradideris, nequicquam Deos implores; irati, infestique sunt.* M. POR. CAT. *in* SALUST.

Printed in the YEAR MDCCXLVII.

Title Page of *Plain Truth*, written in 1747.

of union and discipline for our defense, and promised to propose in a few days an association to be generally signed for that purpose. The pamphlet had a sudden and surprising effect. I was called upon for the instrument of association, and having settled the draft of it with a few friends, I appointed a meeting of the citizens in the large building before mentioned. The house was pretty full; I had prepared a number of printed copies, and provided pens and ink dispersed all over the room. I harangued them a little on the subject, read the paper and explained it, and then distributed the copies, which were eagerly signed, not the least objection being made.

When the company separated, and the papers were collected, we found above twelve hundred hands; and, other copies being dispersed in the country, the subscribers amounted at length to upward of ten thousand. These all furnished themselves as soon as they could with arms, formed themselves into companies and regiments, chose their own officers, and met every week to be instructed in the manual exercise and other parts of military discipline. The women by subscriptions among themselves provided silk colors which they presented to the companies, painted with different devices and mottoes which I supplied.

The officers of the companies composing the Philadelphia regiment being met, chose me for their colonel; but, conceiving myself unfit, I declined that station and recommended Mr. Lawrence, a fine person and man of influence, who was accordingly appointed. I then proposed a lottery to defray the expense of building a battery below the town and furnishing it with cannon. It filled expeditiously and the battery was soon erected, the merlons [1] being framed of logs and filled with earth. We bought some old cannon from Boston, but, these not being sufficient, we wrote to England for more, soliciting at the same time our Proprietaries for some assistance, though without much expectation of obtaining it.

Meanwhile, Colonel Lawrence, William Allen, Abram Taylor, Esqr., and myself were sent to New York by the associators, commissioned to borrow some cannon of Governor Clinton. He at first refused us peremptorily; but at dinner with his council, where there was great drinking of madeira wine, as the custom of that place then was, he softened by degrees and said he would lend us six. After a few more bumpers he advanced to ten; and at length he very good-naturedly conceded eighteen. They were fine cannon, eighteen-pounders, with their carriages, which we soon transported and mounted on our battery, where the associators kept a nightly guard while the war lasted, and among the rest I regularly took my turn of duty there as a common soldier.

My activity in these operations was agreeable to the Governor and Council; they took me into confidence, and I was consulted by them in every measure wherein their concurrence was thought useful to the association. Calling

[1] Ramparts.

in the aid of religion, I proposed to them the proclaiming a fast to promote reformation and implore the blessing of Heaven on our undertaking. They embraced the motion; but, as it was the first fast ever thought of in the Province, the secretary had no precedent from which to draw the proclamation. My education in New England, where a fast is proclaimed every year, was here of some advantage: I drew it in the accustomed style, it was translated into German, printed in both languages, and divulged [1] through the Province. This gave the clergy of the different sects an opportunity of influencing their congregation to join in the association, and it would probably have been general among all but Quakers if the peace had not soon intervened.

It was thought by some of my friends that by my activity in these affairs I should offend that sect and thereby lose my interest in the Assembly of the Province, where they formed a great majority. A young gentleman who had likewise some friends in the House and wished to succeed me as their clerk acquainted me that it was decided to displace me at the next election; and he, therefore, in good will advised me to resign, as more consistent with my honor than being turned out. My answer to him was that I had read or heard of some public man who made it a rule never to ask for an office, and never to refuse one when offered to him. "I approve," says I, "of his rule and will practise it with a small addition; I shall never *ask*,

[1] Circulated.

never *refuse*, nor ever *resign* an office. If they will have my office of clerk to dispose of to another, they shall take it from me. I will not by giving it up lose my right of some time or other making reprisals on my adversaries." I heard, however, no more of this; I was chosen again unanimously as usual at the next election. Possibly, as they disliked my late intimacy with the members of Council who had joined the governors in all the disputes about military preparations with which the House had long been harassed, they might have been pleased if I would voluntarily have left them; but they did not care to displace me on account merely of my zeal for the association, and they could not well give another reason.

"*Embarrassments that the Quakers Suffered . . .*"

INDEED I had some cause to believe that the defense of the country was not disagreeable to any of them provided they were not required to assist in it. And I found that a much greater number of them than I could have imagined, though against offensive war, were clearly for the defensive. Many pamphlets pro and con were published on the subject, and some by good Quakers in favor of defense, which I believed convinced most of their younger people.

A transaction in our fire company gave me some insight into their prevailing sentiments. It had been proposed that we should encourage the scheme for building a battery by laying out the present stock, then about sixty pounds, in tickets of the lottery. By our rules no money could be disposed of till the next meeting after the proposal. The company consisted of thirty members, of which twenty-two were Quakers and eight only of other persuasions. We eight punctually attended the meeting; but though we thought that some of the Quakers would join us, we were by no means sure of a majority. Only one Quaker, Mr. James Morris, appeared to oppose the measure. He expressed much sorrow that it had ever been proposed, as he said Friends [1] were all against it and it would create such discord as might break up the company. We told him that we saw no reason for that; we were the minority, and if Friends were against the measure and outvoted us, we must and should, agreeably to the usage of all societies, submit. When the hour for business arrived it was moved to put the vote; he allowed we might then do it by the rules, but as he could assure us that a number of members intended to be present for the purpose of opposing it, it would be but candid to allow a little time for their appearing.

While we were disputing this, a waiter came to tell me two gentlemen below desired to speak with me. I went down and found they were two of our Quaker members. They told me there were eight of them assembled at a tavern just by, that they were determined to come and

[1] Society of Friends is the official title of the Quakers.

vote with us if there should be occasion, which they hoped would not be the case, and desired we would not call for their assistance if we could do without it, as their voting for such a measure might embroil them with their elders and friends. Being thus secure of a majority, I went up and after a little seeming hesitation agreed to a delay of another hour. This Mr. Morris allowed to be extremely fair. Not one of his opposing friends appeared, at which he expressed great surprise; and at the expiration of the hour we carried the resolution eight to one; and as of the twenty-two Quakers eight were ready to vote with us, and thirteen by their absence manifested that they were not inclined to oppose the measure, I afterward estimated the proportion of Quakers sincerely against defense as one to twenty-one only; for these were all regular members of that Society and in good reputation among them and had due notice of what was proposed at that meeting.

The honorable and learned Mr. Logan, who had always been of that sect, was one who wrote an address to them declaring his approbation of defensive war and supporting his opinion by many strong arguments. He put into my hands sixty pounds to be laid out in lottery tickets for the battery with directions to apply what prizes might be drawn wholly to that service. He told me the following anecdote of his old master, William Penn, respecting defense. He came over from England when a young man with that Proprietary and as his secretary. It was wartime, and their ship was chased by an armed vessel sup-

Autobiography

posed to be an enemy. Their captain prepared for defense, but told William Penn and his company of Quakers that he did not expect their assistance and they might retire into the cabin, which they did, except James Logan, who chose to stay upon deck and was quartered to a gun. The supposed enemy proved a friend, so there was no fighting; but when the secretary went down to communicate the intelligence, William Penn rebuked him severely for staying upon deck and undertaking to assist in defending the vessel, contrary to the principles of Friends, especially as it had not been required by the captain. This reproof being before all the company piqued the secretary, who answered, "I being thy servant, why did thee not order me to come down? But thee was willing enough that I should stay and help to fight the ship when thee thought there was danger."

My being many years in the Assembly, the majority of which were constantly Quakers, gave me frequent opportunities of seeing the embarrassment given them by their principle against war whenever application was made to them by order of the Crown to grant aids for military purposes. They were unwilling to offend government on the one hand by a direct refusal; and their friends, the body of the Quakers, on the other by a compliance contrary to their principles; hence a variety of evasions to avoid complying and modes of disguising the compliance when it became unavoidable. The common mode at last was to grant money under the phrase of its being "for the King's use," and never to inquire how it was applied.

But if the demand was not directly from the Crown that phrase was found not so proper and some other was to be invented. As, when powder was wanting (I think it was for the garrison at Louisburg) and the government of New England solicited a grant of some from Pennsylvania, which was much urged on the House by Governor Thomas, they could not grant money to buy powder because that was an ingredient of war; but they voted an aid to New England of three thousand pounds to be put into the hands of the Governor and appropriated it for the purchasing of bread, flour, wheat, or *other grain.* Some of the Council, desirous of giving the House still further embarrassment, advised the Governor not to accept provision, as not being the thing he had demanded; but he replied, "I shall take the money, for I understand very well their meaning; other grain is gunpowder," which he accordingly bought, and they never objected to it.

It was in allusion to this fact that, when in our fire company we feared the success of our proposal in favor of the lottery, and I had said to my friend Mr. Syng, one of our members, "If we fail, let us move the purchase of a fire-engine with the money; the Quakers can have no objection to that; and then, if you nominate me and I you as a committee for that purpose, we will buy a great gun, which is certainly a *fire-engine.*" "I see," says he, "you have improved by being so long in the Assembly;

your equivocal project would be just a match for their wheat or *other grain.*"

These embarrassments that the Quakers suffered from having established and published it as one of their principles that no kind of war was lawful, and which being once published they could not afterwards, however they might change their minds, easily get rid of, reminds me of what I think a more prudent conduct in another sect among us, that of the Dunkers.[1] I was acquainted with one of its founders, Michael Welfare, soon after it appeared. He complained to me that they were grievously calumniated by the zealots of other persuasions and charged with abominable principles and practices to which they were utter strangers. I told him this had always been the case with new sects and that to put a stop to such abuse I imagined it might be well to publish the articles of their belief and the rules of their discipline. He said that it had been proposed among them but not agreed to for this reason: "When we were first drawn together as a society," says he, "it had pleased God to enlighten our minds so far as to see that some doctrines which we once esteemed truths were errors; and that others which we had esteemed errors were real truths. From time to time He has been pleased to afford us farther light, and our principles have been improving and our errors diminishing. Now we are not sure that we are arrived at the end of this progres-

[1] Also known as Dunkards, from their practice of baptizing by immersion.

sion and at the perfection of spiritual or theological knowledge; and we fear that if we should once print our confession of faith we should feel ourselves as if bound and confined by it and perhaps be unwilling to receive further improvement, and our successors still more so, as conceiving what we their elders and founders had done to be something sacred never to be departed from."

This modesty in a sect is perhaps a singular instance in the history of mankind, every other sect supposing itself in possession of all truth and that those who differ are so far in the wrong; like a man traveling in foggy weather, those at some distance before him on the road he sees wrapped up in the fog as well as those behind him and also the people in the fields on each side, but near him all appears clear, though in truth he is as much in the fog as any of them. To avoid this kind of embarrassment, the Quakers have of late years been gradually declining the public service in the Assembly and in the magistracy, choosing rather to quit their power than their principle.

"An Opportunity to Serve Others . . ."

IN ORDER of time I should have mentioned before that having in 1742 invented an open stove[1] for the better warming of rooms and at the same time saving fuel, as the fresh air admitted was warmed in entering. I made a present of the model to Mr. Robert Grace, one of my

[1] This type of stove still bears Franklin's name.

ADVERTISEMENT.

THESE Fire-Places are made in the best Manner, and sold by *R. Grace* in *Philadelphia*. They are sold also by *J. Parker* in *New-York*, and by *J. Franklin* in *Boston*.

The within-describ'd is of the middle and most common Size: There are others to be had both larger and smaller.

ERRATA. Page 9. line 6, of the Notes, read *Sanitate*. line 20, read *motus*. Page 14, line 21, after the Word *away*, add, *also a square Hole b for the Bellows*.

First pages of *An Account of the New-Invented Pennsylvania Fire-Places*, written by Franklin in 1744.

[1]

AN ACCOUNT

Of the New-Invented

FIRE-PLACES.

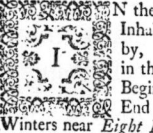

IN these Northern Colonies the Inhabitants keep Fires to sit by, generally *Seven Months* in the Year; that is, from the Beginning of *October* to the End of *April*; and in some Winters near *Eight Months*, by taking in part of *September* and *May*.

Wood, our common Fewel, which within these 100 Years might be had at every Man's Door, must now be fetch'd near 100 Miles to some Towns, and makes a very considerable Article in the Expence of Families.

A

early friends, who having an iron-furnace, found the casting of the plates for these stoves a profitable thing, as they were growing in demand. To promote that demand, I wrote and published a pamphlet entitled *An Account of the New-Invented Pennsylvania Fire-Places; wherein Their Construction and Manner of Operation is Particularly Explained; Their Advantages above Every Other Method of Warming Rooms Demonstrated; and All Objections That have been Raised against the Use of Them Answered and Obviated*, etc. This pamphlet had a good effect. Governor Thomas was so pleased with the construction of this stove, as described in it, that he offered to give me a patent for the sole vending of them for a term of years; but I declined it from a principle which has ever weighed with me on such occasions, viz., That, as we enjoy great advantages from the inventions of others, we should be glad of an opportunity to serve others by any invention of ours; and this we should do freely and generously.

An ironmonger in London, however, assuming [1] a good deal of my pamphlet and working it up into his own and making some small changes in the machine, which rather hurt its operation, got a patent for it there, and made, as I was told, a little fortune by it. And this is not the only instance of patents taken out for my inventions by others, though not always with the same success, which I never contested, as having no desire of profiting by patents my-

[1] Appropriating without permission.

self and hating disputes. The use of these fireplaces in very many houses both of this and the neighboring colonies has been, and is, a great saving of wood to the inhabitants.

Peace being concluded and the association business therefore at an end, I turned my thoughts again to the affair of establishing an academy. The first step I took was to associate in the design a number of active friends, of whom the Junto furnished a good part; the next was to write and publish a pamphlet entitled *Proposals Relating to the Education of Youth in Pennsylvania*. This I distributed among the principal inhabitants gratis; and as soon as I could suppose their minds a little prepared by the perusal of it, I set on foot a subscription for opening and supporting an academy; it was to be paid in quotas yearly for five years; by so dividing it I judged the subscription might be larger, and I believe it was so, amounting to no less, if I remember right, than five thousand pounds.

In the introduction to these proposals I stated their publication, not as an act of mine, but of some *public-spirited gentleman*, avoiding as much as I could, according to my usual rule, the presenting myself to the public as the author of any scheme for their benefit.

The subscribers to carry the project into immediate execution chose out of their number twenty-four trustees and appointed Mr. Francis, then attorney-general, and myself to draw up constitutions for the government of the academy; which being done and signed, a house was hired, masters engaged, and the schools opened, I think, in the same year, 1749.

The scholars increasing fast, the house was soon found too small, and we were looking out for a piece of ground, properly situated, with intention to build, when Providence threw into our way a large house ready built, which with a few alterations might well serve our purpose. This was the building before mentioned, erected by the hearers of Mr. Whitefield, and was obtained for us in the following manner.

It is to be noted that the contributions to this building being made by people of different sects, care was taken in the nomination of trustees in whom the building and ground was to be vested that a predominancy should not be given to any sect lest in time that predominancy might be a means of appropriating the whole to the use of such sect contrary to the original intention. It was therefore that one of each sect was appointed, viz., one Church-of-England man, one Presbyterian, one Baptist, one Moravian,[1] etc., those, in case of vacancy by death, were to fill it by election from among the contributors. The Moravian happened not to please his colleagues, and on his death they resolved to have no other of that sect. The difficulty then was how to avoid having two of some other sect by means of the new choice.

Several persons were named and for that reason not

[1] A Christian sect in southern Pennsylvania tracing their origin to John Huss. The Scripture is their sole rule of faith and practice.

Autobiography

agreed to. At length one mentioned me, with the observation that I was merely an honest man and of no sect at all, which prevailed with them to choose me. The enthusiasm which existed when the house was built had long since abated, and its trustees had not been able to procure fresh contributions for paying the ground-rent and discharging some other debts the building had occasioned, which embarrassed them greatly. Being now a member of both sets of trustees, that for the building and that for the academy, I had a good opportunity of negotiating with both and brought them finally to an agreement by which the trustees for the building were to cede it to those of the academy, the latter undertaking to discharge the debt, to keep forever open in the building a large hall for occasional preachers according to the original intention, and maintain a free school for the instruction of poor children. Writings were accordingly drawn and on paying the debts the trustees of the academy were put in possession of the premises; and by dividing the great and lofty hall into stories and different rooms above and below for the several schools and purchasing some additional ground, the whole was soon made fit for our purpose and the scholars removed into the building. The care and trouble of agreeing with the workmen, purchasing materials, and superintending the work, fell upon me; and I went through it the more cheerfully as it did not then interfere with my private business, having the year before taken a very able, industrious, and honest partner, Mr. David

Hall, with whose character I was well acquainted, as he had worked for me four years. He took off my hands all care of the printing-office, paying me punctually my share of the profits. The partnership continued eighteen years, successfully for us both.

The trustees of the academy after a while were incorporated by a charter from the Governor; their funds were increased by contributions in Britain and grants of land from the Proprietaries, to which the Assembly has since made considerable addition; and thus was established the present University of Philadelphia.[1] I have been continued one of its trustees from the beginning, now near forty years, and have had the very great pleasure of seeing a number of the youth who have received their education in it distinguished by their improved abilities, serviceable in public stations, and ornaments to their country.

"*The Public Laid Hold of Me for Their Purposes . . .*"

WHEN I disengaged myself, as above mentioned, from private business, I flattered myself that by the sufficient though moderate fortune I had acquired I had secured leisure during the rest of my life for philosophical studies and amusements. I purchased all Dr. Spence's apparatus, who had come from England to lecture here, and I proceeded in my electrical experiments with great alacrity;

[1] Now the University of Pennsylvania.

but the public, now considering me as a man of leisure, laid hold of me for their purposes,[1] every part of our civil government, and almost at the same time, imposing some duty upon me. The Governor put me into the commission of the peace; the corporation of the city chose me of the common council and soon after an alderman; and the citizens at large chose me a burgess to represent them in Assembly. This latter station was the more agreeable to me as I was at length tired with sitting there to hear debates in which as clerk I could take no part, and which were often so unentertaining that I was induced to amuse myself with making magic squares or circles or anything to avoid weariness; and I conceived my becoming a member would enlarge my power of doing good. I would not, however, insinuate that my ambition was not flattered by all these promotions; it certainly was; for considering my low beginning they were great things to me; and they were still more pleasing as being so many spontaneous testimonies of the public good opinion and by me entirely unsolicited.

The office of justice of the peace I tried a little by attending a few courts and sitting on the bench to hear causes; but finding that more knowledge of the common law than I possessed was necessary to act in that station with credit, I gradually withdrew from it, excusing myself by my being obliged to attend the higher duties of a legis-

[1] Though every edition reads thus, it is apparent that a line has been omitted in copying the manuscript.

A

TREATY, &c.

To the Honourable JAMES HAMILTON, *Esq; Lieutenant-Governor, and Commander in Chief, of the Province of* Pennsylvania, *and Counties of* New-Castle, Kent *and* Sussex, *upon* Delaware,

The REPORT *of* Richard Peters, Isaac Norris, *and* Benjamin Franklin, *Esquires, Commissioners appointed to treat with some Chiefs of the* Ohio Indians, *at* Carlisle, *in the County of* Cumberland, *by a Commission, bearing Date the* 22d *Day of* September, 1753.

May it please the Governor,

NOT knowing but the *Indians* might be waiting at *Carlisle,* we made all the Dispatch possible, as soon as we had received our Commission, and arrived there on the Twenty-sixth, but were agreeably surprized to find that they came there only that Day. 1753.

Immediately on our Arrival we conferred with *Andrew Montour,* and *George Croghan,* in order to know from them what had occasioned the present coming of the *Indians,* that we might, by their Intelligence, regulate our first Intercourse with them; and were informed, that tho' their principal Design, when they left *Ohio,* was to hold a Treaty with the Government of *Virginia,* at *Winchester,* where they had accordingly been; yet they intended a Visit to this Province, to which they had been frequently encouraged by *Andrew Montour,* who told them, he had the Governor's repeated Orders to invite them to come and see him, and assured them of an hearty Welcome; and that they had moreover some important Matters to propose and transact with this Government.

The Commissioners finding this to be the Case, and that these *Indians* were some of the most considerable Persons of the *Six Nations, Delawares, Shawonese,* with Deputies from the *Twightwees,* and *Owendaets,* met them in Council, in which the Commissioners declared the Contents of their Commission, acknowledged the Governor's Invitation, and bid them heartily welcome among their Brethren of *Pennsylvania,* to whom their Visit was extremely agreeable.---*Conrad Weiser* and *Andrew Montour* interpreting between the Commissioners and *Indians,* and several Magistrates, and others, of the principal Inhabitants of the County, favouring them with their Presence.

The *Twightwees* and *Delawares* having had several of their great Men cut off by the *French* and their *Indians,* and all the Chiefs of the *Owendaets* being lately dead, it became necessary to condole their Loss; and no Business could be begun, agreeable to the *Indian* Customs, till the Condolances were passed; and as these could not be made, with the usual Ceremonies, for want of the Goods, which were not arrived, and it was uncertain when they would, the Commissioners were put to some Difficulties, and ordered the Interpreters to apply to *Starrooyady,* an *Oneido* Chief; who had the Conduct of the Treaty in *Virginia,* and was a Person of great Weight in their Councils, and to ask his Opinion, whether the Condolances would be accepted by Belts and Strings, and Lists of the particular Goods intended to be given, with Assurances of their Delivery as soon as they should come. *Scarrooyady* was pleased with the Application; but frankly declared, that the *Indians* could not proceed to Business while the Blood remained on their Garments, and that the Condolances could not be accepted unless the Goods, intended to cover the Graves, were actually spread on the Ground before them. A Messenger was therefore forthwith sent to meet and hasten the Waggoners, since every Thing must stop till the Goods came.

It was then agreed to confer with *Starrooyady,* and some other of the Chiefs of the *Shawonese* and *Delawares,* on the State of Affairs at *Ohio,* and from them the Commissioners learned, in sundry Conferences, the following Particulars, *viz.*

" That when the Governor of *Pennsylvania's* Express arrived at *Ohio,* with the Account of the March of a large *French* Army to the Heads of *Ohio,* with Intent to take Possession of that Country, it alarmed the *Indians* so much, that the *Delawares,* at *Weningo,* an *Indian* Town, situate high up on *Ohio* River, went, agreeable to a Custom established among the *Indians,* and forbad, by a formal Notice, the Commander of that Armament, then advanced to the *Straits,* between Lake *Ontario* and Lake *Erie,* to continue his March, at least not to presume to come farther than *Niagara :* This had not

First page of *A Treaty Held with the Ohio Indians at Carlisle in October, 1753.*

Autobiography

lator in the Assembly. My election to this trust was repeated every year for ten years without my ever asking any elector for his vote or signifying either directly or indirectly any desire of being chosen. On taking my seat in the House, my son was appointed their clerk.

The year following, a treaty being to be held with the Indians at Carlisle, the Governor sent a message to the House proposing that they should nominate some of their members to be joined with some members of Council as commissioners for that purpose. The House named the speaker (Mr. Norris) and myself; and, being commissioned, we went to Carlisle and met the Indians accordingly.

As those people are extremely apt to get drunk and, when so, are very quarrelsome and disorderly, we strictly forbade the selling any liquor to them; and when they complained of this restriction, we told them that if they would continue sober during the treaty, we would give them plenty of rum when business was over. They promised this, and they kept their promise because they could get no liquor, and the treaty was conducted very orderly and concluded to mutual satisfaction. They then claimed and received the rum; this was in the afternoon: they were near one hundred men, women, and children, and were lodged in temporary cabins, built in the form of a square, just without the town. In the evening, hearing a great noise among them, the commissioners walked out to see what was the matter. We found they had made a great bonfire in the middle of the square; they were all drunk, men and women, quarreling and fighting. Their dark-colored bodies, half naked, seen only by the gloomy light of the bonfire, running after and beating one another with firebrands, accompanied by their horrid yellings, formed a scene the most resembling our ideas of hell that could well be imagined; there was no appeasing the tumult, and we retired to our lodging. At midnight a number of them came thundering at our door, demanding more rum, of which we took no notice.

The next day, sensible they had misbehaved in giving us that disturbance, they sent three of their old counselors to make their apology. The orator acknowledged the fault, but laid it upon the rum; and then endeavored to excuse the rum by saying, "The Great Spirit who made all things made everything for some use, and whatever use he designed anything for, that use it should always be put to. Now when he made rum he said, 'Let this be for the Indians to get drunk with,' and it must be so." And, indeed, if it be the design of Providence to extirpate these savages in order to make room for cultivators of the earth, it seems not improbable that rum may be the appointed means. It has already annihilated all the tribes who formerly inhabited the seacoast.

In 1751 Dr. Thomas Bond, a particular friend of mine, conceived the idea of establishing a hospital in Philadelphia (a very beneficent design, which has been ascribed to me, but was originally his) for the reception and cure of poor sick persons, whether inhabitants of the Province or strangers. He was zealous and active in endeavoring to procure subscriptions for it, but the proposal being a novelty in America and at first not well understood, he met with but small success.

At length he came to me with the compliment that he found there was no such thing as carrying a public-spirited project through without my being concerned in it. "For," says he, "I am often asked by those to whom I propose subscribing, 'Have you consulted Franklin upon this business? And what does he think of it?' And when I tell them that I have not (supposing it rather out of your line), they do not subscribe, but say they will consider of it." I inquired into the nature and probable utility of his scheme, and receiving from him a very satisfactory explanation, I not only subscribed to it myself, but engaged heartily in the design of procuring subscriptions from others. Previously, however, to the solicitation, I endeavored to prepare the minds of the people by writing on the subject in the newspapers, which was my usual custom in such cases, but which he had omitted.

The subscriptions afterwards were more free and generous, but, beginning to flag, I saw they would be insufficient without some assistance from the Assembly, and therefore proposed to petition for it, which was done. The country members did not at first relish the project; they objected that it could only be serviceable to the city, and therefore the citizens alone should be at the expense of it; and they doubted whether the citizens themselves generally approved of it. My allegation on the contrary, that it met with such approbation as to leave no doubt of our being able to raise two thousand pounds by voluntary donations, they considered as a most extravagant supposition and utterly impossible.

On this I formed my plan; and, asking leave to bring in a bill for incorporating the contributors according to the prayer of their petition and granting them a blank sum of money, which leave was obtained chiefly on the consideration that the House could throw the bill out if they did not like it, I drew it so as to make the important clause a conditional one, viz., "And be it enacted by the authority aforesaid that when the said contributors shall have met and chosen their managers and treasurer *and shall have raised by their contributions a capital stock of —— value* (the yearly interest of which is to be applied to the accommodating of the sick poor in the said hospital free of charge for diet, attendance, advice, and medicines), *and shall make the same appear to the satisfaction of the speaker of the Assembly for the time being*, that *then* it shall and may be lawful for the said speaker, and he is hereby required, to sign an order on the provincial treasurer for the payment of two thousand pounds in two yearly payments to the treasurer of the said hospital, to be applied to the founding, building, and finishing of the same."

This condition carried the bill through; for the members who had opposed the grant and now conceived they might have the credit of being charitable without the expense agreed to its passage; and then in soliciting subscriptions among the people we urged the conditional promise of the law as an additional motive to give, since every man's donation would be doubled; thus the clause worked both ways. The subscriptions accordingly soon exceeded the requisite sum, and we claimed and received the public gift, which enabled us to carry the design into execution. A convenient and handsome building was soon erected; the institution has by constant experience been found useful and flourishes to this day; and I do not remember any of my political maneuvers the success of which gave me at the time more pleasure, or wherein after thinking of it I more easily excused myself for having made some use of cunning.

It was about this time that another projector, the Rev. Gilbert Tennent, came to me with a request that I would assist him in procuring a subscription for erecting a new meeting-house. It was to be for the use of a congregation he had gathered among the Presbyterians who were originally disciples of Mr. Whitefield. Unwilling to make myself disagreeable to my fellow-citizens by too frequently soliciting their contributions, I absolutely refused. He then desired I would furnish him with a list of the names of persons I knew by experience to be generous and public-spirited. I thought it would be unbecoming in me after their kind compliance with my solicitations to mark them out to be worried by other beggars, and therefore refused also to give such a list. He then desired I would at least give him my advice. "That I will readily do," said I; "and in the first place I advise you to apply to all those who you know will give something; next, to those who you are uncertain whether they will give anything or not and show them the list of those who have given; and lastly do not neglect those who you are sure will give nothing, for in some of them you may be mistaken." He laughed and thanked me, and said he would take my advice. He did so, for he asked of *everybody*, and he obtained a much larger sum than he expected with which he erected the capacious and very elegant meeting-house that stands in Arch Street.

"Paving the City . . ."

OUR CITY, though laid out with a beautiful regularity, the streets large, straight, and crossing each other at right angles, had the disgrace of suffering those streets to remain long unpaved, and in wet weather the wheels of heavy carriages ploughed them into a quagmire so that it was difficult to cross them, and in dry weather the dust was offensive. I had lived near what was called the Jersey Market, and saw with pain the inhabitants wading in mud while purchasing their provisions. A strip of ground down the middle of that market was at length paved with brick so that being once in the market they had firm footing, but were often over shoes in dirt to get there. By talking and writing on the subject I was at length instrumental in getting the street paved with stone between the market and the bricked foot pavement that was on each side next the houses. This for some time gave an easy access to the market dry-shod; but, the rest of the street not being paved, whenever a carriage came out of the mud upon this pavement, it shook off and left its dirt upon it, and it was soon covered with mire which was not removed, the city as yet having no scavengers.[1]

After some inquiry, I found a poor, industrious man, who was willing to undertake keeping the pavement clean by sweeping it twice a week, carrying off the dirt from before all the neighbors' doors for the sum of sixpence per month to be paid by each house. I then wrote and printed a paper setting forth the advantages to the neighborhood that might be obtained by this small expense; the greater ease in keeping our houses clean, so much dirt not being brought in by people's feet; the benefit to the shops by more custom, etc., etc., as buyers could more easily get at them; and by not having, in windy weather, the dust blown in upon their goods, etc., etc. I sent one of these papers to each house, and in a day or two went round to see who would subscribe an agreement to pay these sixpences; it was unanimously signed and for a time well executed. All the inhabitants of the city were delighted with the cleanliness of the pavement that sur-

[1] Street cleaners.

rounded the market, it being a convenience to all, and this raised a general desire to have all the streets paved and made the people more willing to submit to a tax for that purpose.

After some time I drew a bill for paving the city, and brought it into the Assembly. It was just before I went to England in 1757, and did not pass till I was gone, and then with an alteration in the mode of assessment which I thought not for the better, but with an additional provision for lighting as well as paving the streets which was a great improvement. It was by a private person, the late Mr. John Clifton, his giving a sample of the utility of lamps, by placing one at his door, that the people were first impressed with the idea of enlighting all the city. The honor of this public benefit has also been ascribed to me, but it belongs truly to that gentleman. I did but follow his example and have only some merit to claim respecting the form of our lamps as differing from the globe lamps we were at first supplied with from London. Those we found inconvenient in these respects: they admitted no air below; the smoke, therefore, did not readily go out above, but circulated in the globe, lodged on its inside, and soon obstructed the light they were intended to afford, giving besides the daily trouble of wiping them clean; and an accidental stroke on one of them would demolish it and render it totally useless. I therefore suggested the composing them of four flat panes with a long funnel above to draw up the smoke and crevices admitting air below to facilitate the ascent of the smoke; by this means they were kept clean and did not grow dark in a few hours as the London lamps do, but continued bright till morning, and an accidental stroke would generally break but a single pane, easily repaired.

I have sometimes wondered that the Londoners did not, from the effect holes in the bottom of the globe lamps used at Vauxhall [1] have in keeping them clean, learn to have such holes in their street lamps. But these holes being made for another purpose, viz., to communicate flame more suddenly to the wick by a little flax hanging down through them, the other use, of letting in air, seems not to have been thought of; and therefore after the lamps have been lit a few hours, the streets of London are very poorly illuminated.

The mention of these improvements puts me in mind of one I proposed when in London to Dr. Fothergill, who was among the best men I have known and a great promoter of useful projects. I had observed that the streets, when dry, were never swept, and the light dust carried away; but it was suffered to accumulate till wet weather reduced it to mud, and then after lying some days so deep on the pavement that there was no crossing but in paths kept clean by poor people with brooms, it was with great labor raked together and thrown up into carts open above, the sides of which suffered some of the slush at every jolt on the pavement to shake out and fall, some-

[1] An amusement park in London.

times to the annoyance of foot-passengers. The reason given for not sweeping the dusty streets was that the dust would fly into the windows of shops and houses.

An accidental occurrence had instructed me how much sweeping might be done in a little time. I found at my door in Craven Street one morning a poor woman sweeping my pavement with a birch broom; she appeared very pale and feeble, as just come out of a fit of sickness. I asked who employed her to sweep there; she said, "Nobody, but I am very poor and in distress, and I sweeps before gentlefolks's doors and hopes they will give me something." I bid her sweep the whole street clean, and I would give her a shilling; this was at nine o'clock; at twelve she came for the shilling. From the slowness I saw at first in her working, I could scarce believe that the work was done so soon and sent my servant to examine it, who reported that the whole street was swept perfectly clean, and all the dust placed in the gutter, which was in the middle; and the next rain washed it quite away, so that the pavement and even the kennel [1] were perfectly clean.

"Keeping Clean the Streets of London . . ."

I THEN judged that if that feeble woman could sweep such a street in three hours, a strong, active man might have done it in half the time. And here let me remark the convenience of having but one gutter in such a narrow

[1] Gutter.

street, running down its middle, instead of two, one on each side, near the footway; for where all the rain that falls on a street runs from the sides and meets in the middle, it forms there a current strong enough to wash away all the mud it meets with; but when divided into two channels it is often too weak to cleanse either and only makes the mud it finds more fluid so that the wheels of carriages and feet of horses throw and dash it upon the foot-pavement, which is thereby rendered foul and slippery, and sometimes splash it upon those who are walking. My proposal, communicated to the good doctor, was as follows:

"For the more effectual cleaning and keeping clean the streets of London and Westminster, it is proposed that the several watchmen be contracted with to have the dust swept up in dry seasons and the mud raked up at other times, each in the several streets and lanes of his round; that they be furnished with brooms and other proper instruments for these purposes to be kept at their respective stands, ready to furnish the poor people they may employ in the service.

"That in the dry summer months the dust be all swept up into heaps at proper distances before the shops and windows of houses are usually opened, when the scavengers with close-covered carts shall also carry it all away.

"That the mud when raked up be not left in heaps to be spread abroad again by the wheels of carriages and trampling of horses, but that the scavengers be provided with bodies of carts, not placed high upon wheels, but low upon sliders, with lattice bottoms, which, being covered with straw, will retain the mud thrown into them and permit the water to drain from it, whereby it will become much lighter, water making the greatest part of its weight; these bodies of carts to be placed at convenient distances and the mud brought to them in wheelbarrows, they remaining where placed till the mud is drained, and then horses brought to draw them away."

I have since had doubts of the practicability of the latter part of this proposal on account of the narrowness of some streets and the difficulty of placing the draining-sleds so as not to encumber too much the passage; but I am still of opinion that the former, requiring the dust to be swept up and carried away before the shops are open, is very practicable in the summer, when the days are long; for in walking through the Strand and Fleet Street one morning at seven o'clock I observed there was not one shop open, though it had been daylight and the sun up above three hours, the inhabitants of London choosing voluntarily to live much by candle-light and sleep by sunshine, and yet often complain, a little absurdly, of the duty on candles and the high price of tallow.

Some may think these trifling matters not worth minding or relating; but when they consider that though dust blown into the eyes of a single person or into a single shop on a windy day is but of small importance, yet the great number of the instances in a populous city and its frequent repetitions give it weight and consequence, perhaps they will not censure very severely those who bestow some attention to affairs of this seemingly low nature. Human felicity is produced not so much by great pieces of good fortune that seldom happen as by little advantages that occur every day. Thus if you teach a poor young man to shave himself and keep his razor in order, you may contribute more to the happiness of his life than in giving him a thousand guineas. The money may be soon spent, the regret only remaining of having foolishly consumed it; but in the other case he escapes the frequent vexation of waiting for barbers and of their sometimes dirty fingers, offensive breaths, and dull razors; he shaves when most convenient to him and enjoys daily the pleasure of its being done with a good instrument. With these sentiments I have hazarded the few preceding pages, hoping they may afford hints which some time or other may be useful to a city I love, having lived many years in it very happily, and perhaps to some of our towns in America.

Having been for some time employed by the postmaster-general of America as his comptroller in regulating several offices and bringing the officers to account, I was upon his death in 1753 appointed jointly with Mr. William Hunter to succeed him, by a commission from the postmaster-general in England. The American office never had hitherto paid anything to that of Britain. We were to have six hundred pounds a year between us if we could make that sum out of the profits of the office. To do this a variety of improvements were necessary; some of these were inevitably at first expensive, so that in the first four years the office became above nine hundred pounds in debt to us. But it soon after began to repay us; and before I was displaced by a freak of the ministers, of which I shall speak hereafter, we had brought it to yield *three times* as much clear revenue to the Crown as the post office of Ireland. Since that imprudent transaction, they have received from it—not one farthing!

The business of the post office occasioned my taking a journey this year to New England, where the College of Cambridge of their own motion presented me with the degree of Master of Arts. Yale College in Connecticut had before made me a similar compliment.[1] Thus without studying in any college I came to partake of their honors. They were conferred in consideration of my improvements and discoveries in the electric branch of natural philosophy.

"*A Plan for the Union of all the Colonies . . .*"

IN 1754, war with France being again apprehended, a congress of commissioners from the different colonies

[1] Harvard gave Franklin the Master of Arts degree in July, Yale in September, 1753, and William and Mary in April, 1756. Oxford made him Doctor of Civil Law in April, 1762.

Autobiography

was by an order of the Lords of Trade to be assembled at Albany, there to confer with the chiefs of the Six Nations concerning the means of defending both their country and ours. Governor Hamilton, having received this order, acquainted the House with it, requesting they would furnish proper presents for the Indians to be given on this occasion, and naming the speaker (Mr. Norris) and myself to join Mr. Thomas Penn and Mr. Secretary Peters as commissioners to act for Pennsylvania. The House approved the nomination and provided the goods for the present, though they did not much like treating out of the provinces, and we met the other commissioners at Albany about the middle of June.

In our way thither I projected and drew a plan for the union of all the colonies under one government so far as might be necessary for defense and other important general purposes. As we passed through New York, I had there shown my project to Mr. James Alexander and Mr. Kennedy, two gentlemen of great knowledge in public affairs, and, being fortified by their approbation, I ventured to lay it before the Congress. It then appeared that several of the commissioners had formed plans of the same kind. A previous question was first taken whether a union should be established, which passed in the affirmative unanimously. A committee was then appointed, one member from each colony, to consider the several plans and report. Mine happened to be preferred, and with a few amendments was accordingly reported.

By this plan the general government was to be administered by a president-general, appointed and supported by the Crown, and a grand council was to be chosen by the representatives of the people of the several colonies met in their respective assemblies. The debates upon it in Congress went on daily, hand in hand with the Indian business. Many objections and difficulties were started, but at length they were all overcome and the plan was unanimously agreed to, and copies ordered to be transmitted to the Board of Trade and to the assemblies of the several provinces. Its fate was singular: the assemblies did not adopt it, as they all thought there was too much *prerogative* [1] in it, and in England it was judged to have too much of the *democratic*. The Board of Trade therefore did not approve of it nor recommend it for the approbation of His Majesty; but another scheme was formed, supposed to answer the same purpose better, whereby the governors of the provinces with some members of their respective councils were to meet and order the raising of troops, building of forts, etc., and to draw on the treasury of Great Britain for the expense, which was afterwards to be refunded by an act of Parliament laying a tax on America. My plan, with my reasons in support of it, is to be found among my political papers that are printed.

Being the winter following in Boston, I had much conversation with Governor Shirley upon both the plans. Part of what passed between us on the occasion may also

[1] Royal authority.

be seen among those papers. The different and contrary reasons of dislike to my plan makes me suspect that it was really the true medium; and I am still of opinion it would have been happy for both sides the water if it had been adopted. The Colonies, so united, would have been sufficiently strong to have defended themselves; there would then have been no need of troops from England; of course, the subsequent pretense for taxing America and the bloody contest it occasioned would have been avoided. But such mistakes are not new; history is full of the errors of states and princes.

> *Look round the habitable world, how few*
> *Know their own good, or, knowing it, pursue!*

Those who govern, having much business on their hands, do not generally like to take the trouble of considering and carrying into execution new projects. The best public measures are therefore seldom *adopted from previous wisdom, but forced by the occasion.*

The Governor of Pennsylvania in sending it down to the Assembly expressed his approbation of the plan, "as appearing to him to be drawn up with great clearness and strength of judgment, and therefore recommended it as well worthy of their closest and most serious attention." The House, however, by the management of a certain member, took it up when I happened to be absent, which I thought not very fair, and reprobated it without paying any attention to it at all, to my no small mortification.

"Our New Governor, Mr. Morris . . ."

IN MY journey to Boston this year, I met at New York with our new Governor, Mr. Morris, just arrived there from England, with whom I had been before intimately acquainted. He brought a commission to supersede Mr. Hamilton, who, tired with the disputes his proprietary instructions subjected him to, had resigned. Mr. Morris asked me if I thought he must expect as uncomfortable an administration. I said, "No; you may, on the contrary, have a very comfortable one if you will only take care not to enter into any dispute with the Assembly." "My dear friend," says he, pleasantly, "how can you advise my avoiding disputes? You know I love disputing; it is one of my greatest pleasures; however, to show the regard I have for your counsel, I promise you I will if possible avoid them." He had some reason for loving to dispute, being eloquent, an acute sophister, and therefore generally successful in argumentative conversation. He had been brought up to it from a boy, his father, as I have heard, accustoming his children to dispute with one another for his diversion while sitting at table after dinner; but I think the practice was not wise, for in the course of my observation these disputing, contradicting, and confuting people are generally unfortunate in their affairs. They get victory sometimes, but they never get good will, which would be of more use to them. We parted, he going to Philadelphia, and I to Boston.

In returning, I met at New York with the votes of the Assembly, by which it appeared that, notwithstanding his promise to me, he and the House were already in high contention; and it was a continual battle between them as long as he retained the government. I had my share of it; for as soon as I got back to my seat in the Assembly, I was put on every committee for answering his speeches and messages, and by the committees always desired to make the drafts. Our answers, as well as his messages, were often tart, and sometimes indecently abusive; and as he knew I wrote for the Assembly, one might have imagined that when we met we could hardly avoid cutting throats; but he was so good-natured a man that no personal difference between him and me was occasioned by the contest, and we often dined together.

One afternoon in the height of this public quarrel we met in the street. "Franklin," says he, "you must go home with me and spend the evening; I am to have some company that you will like"; and, taking me by the arm, he led me to his house. In gay conversation over our wine after supper he told us jokingly that he much admired the idea of Sancho Panza, who, when it was proposed to give him a government, requested it might be a government of *blacks*, as then, if he could not agree with his people, he might sell them. One of his friends, who sat next to me, says, "Franklin, why do you continue to side with these damned Quakers? Had not you better sell them? The Proprietor would give you a good price." "The Governor," says I, "has not yet *blacked* them enough." He, indeed, had labored hard to blacken the Assembly in all his messages, but they wiped off his coloring as fast as he laid it on and placed it in return thick upon his own face; so that finding he was likely to be negrofied himself, he as well as Mr. Hamilton grew tired of the contest and quitted the government.

These public quarrels were all at bottom owing to the Proprietaries, our hereditary governors, who, when any expense was to be incurred for the defense of their province, with incredible meanness instructed their deputies to pass no act for levying the necessary taxes unless their vast estates were in the same act expressly excused; and they had even taken bonds of these deputies to observe such instructions. The Assemblies for three years held out against this injustice, though constrained to bend at last. At length Captain Denny, who was Governor Morris's successor, ventured to disobey those instructions; how that was brought about I shall show hereafter.

But I am got forward too fast with my story: there are still some transactions to be mentioned that happened during the administration of Governor Morris.

"War Being in a Manner Commenced with France . . ."

WAR [1] BEING in a manner commenced with France, the government of Massachusetts Bay projected an attack upon Crown Point and sent Mr. Quincy to Pennsylvania, and Mr. Pownall, afterward Governor Pownall, to New York to solicit assistance. As I was in the Assembly, knew its temper, and was Mr. Quincy's countryman, he applied to me for my influence and assistance. I dictated his address to them, which was well received. They voted an aid of ten thousand pounds to be laid out in provisions. But the Governor refusing his assent to their bill (which included this with other sums granted for the use of the Crown) unless a clause were inserted exempting the proprietary estate from bearing any part of the tax that would be necessary, the Assembly, though very desirious of making their grant to New England effectual, were at a loss how to accomplish it. Mr. Quincy labored hard with the Governor to obtain his assent, but he was obstinate.

I then suggested a method of doing the business without the Governor by orders on the trustees of the Loan Office, which by law the Assembly had the right of drawing. There was indeed little or no money at that time in the office, and therefore I proposed that the orders should be

[1] War between England and France known in Europe as the Seven Years' War and in the United States as the French and Indian War.

payable in a year and to bear an interest of five per cent. With these orders I supposed the provisions might easily be purchased. The Assembly with very little hesitation adopted the proposal. The orders were immediately printed, and I was one of the committee directed to sign and dispose of them. The fund for paying them was the interest of all the paper currency then extant in the Province upon loan, together with the revenue arising from the excise,[1] which being known to be more than sufficient, they obtained instant credit and were not only received in payment for the provisions, but many moneyed people who had cash lying by them vested it in those orders, which they found advantageous, as they bore interest while upon hand and might on any occasion be used as money, so that they were eagerly all bought up and in a few weeks none of them were to be seen. Thus this important affair was by my means completed. Mr. Quincy returned thanks to the Assembly in a handsome memorial, went home highly pleased with the success of his embassy, and ever after bore for me the most cordial and affectionate friendship.

The British government, not choosing to permit the union of the Colonies as proposed at Albany and to trust that union with their defense lest they should thereby grow too military and feel their own strength, suspicions and jealousies at this time being entertained of them, sent over General Braddock with two regiments of regular

[1] internal tax on tobacco, liquor, etc.

Autobiography 213

English troops for that purpose. He landed at Alexandria in Virginia and thence marched to Fredericktown [1] in Maryland, where he halted for carriages. Our Assembly apprehending from some information that he had conceived violent prejudices against them as averse to the service, wished me to wait upon him, not as from them, but as postmaster-general, under the guise of proposing to settle with him the mode of conducting with most celerity and certainty the dispatches between him and the governors of the several provinces with whom he must necessarily have continual correspondence and of which they proposed to pay the expense. My son accompanied me on this journey.

We found the General at Fredericktown waiting impatiently for the return of those he had sent through the back parts of Maryland and Virginia to collect wagons. I stayed with him several days, dined with him daily, and had full opportunity of removing all his prejudices by the information of what the Assembly had before his arrival actually done and were still willing to do to facilitate his operations. When I was about to depart, the returns of wagons to be obtained were brought in, by which it appeared that they amounted only to twenty-five, and not all of those were in serviceable condition. The General and all the officers were surprised, declared the expedition was then at an end, being impossible, and exclaimed against the ministers for ignorantly landing them in a

[1] Now Frederick.

country destitute of the means of conveying their stores, baggage, etc., not less than one hundred and fifty wagons being necessary.

I happened to say I thought it was pity they had not been landed rather in Pennsylvania, as in that country almost every farmer had his wagon. The General eagerly laid hold of my words and said, "Then you, sir, who are a man of interest there, can probably procure them for us; and I beg you will undertake it." I asked what terms were to be offered the owners of the wagons, and I was desired to put on paper the terms that appeared to me necessary. This I did, and they were agreed to, and a commission and instructions accordingly prepared immediately. What those terms were will appear in the advertisement I published as soon as I arrived at Lancaster, which being from the great and sudden effect it produced a piece of some curiosity, I shall insert it at length as follows.

"ADVERTISEMENT

"LANCASTER, April 26, 1755

"Whereas, one hundred and fifty wagons, with four horses to each wagon, and fifteen hundred saddle or pack horses are wanted for the service of His Majesty's forces now about to rendezvous at Will's Creek, and his excelency General Braddock having been pleased to empower me to contract for the hire of the same, I hereby give notice that I shall attend for that purpose at Lancaster

ADVERTISEMENT

Lancaster, May 6th. 1755.

NOTICE is hereby given to all who have contracted to send Waggons and Teams, or single Horses from *York* County to the Army at *Will's* Creek, that *David M'Conaughy* and *Michael Schwoope* of the said County, Gentlemen, will attend on my Behalf at *York* Town on *Friday* next, and at *Philip Forney's* on *Saturday*, to value or appraise all such Waggons, Teams and Horses, as shall appear at those Places on the said Days for that Purpose; and such as do not then appear must be valued at *Will's* Creek.

The Waggons that are valued at *York* and *Forney's*, are to set out immediately after the Valuation from thence for *Will's* Creek, under the Conduct and Direction of Persons I shall appoint for that Purpose.

The Owner or Owners of each Waggon or Set of Horses, should bring with them to the Place of Valuation, and deliver to the Appraisers, a Paper containing a Description of their several Horses in Writing, with their several Marks natural and artificial; which Paper is to be annexed to the Contract.

Each Waggon should be furnished with a Cover, that the Goods laden therein may be kept from Damage by the Rain, and the Health of the Drivers preserved, who are to lodge in the Waggons. And each Cover should be marked with the Contractor's Name in large Characters.

Each Waggon, and every Horse Driver should also be furnished with a Hook or Sickle, fit to cut the long Grass that grows in the Country beyond the Mountains.

As all the Waggons are obliged to carry a Load of Oats, or Indian Corn, Persons who have such Grain to dispose of, are desired to be cautious how they hinder the King's Service, by demanding an extravagant Price on this Occasion.

B. FRANKLIN.

Bekantmachung.

Lancaster, 6ten May, 1755.

ES wird hiermit allen u. jeden in York County bekant gemacht, welche accordiret haben Wagen Wagen-Pferdte, oder einzle Pferde, nach der Armee zu Wills's Creek zu schicken, daß David McConaughy und Michael Schoope in besagtem County, meinet wegen den nechsten Freytag zu Yorktown, u. den samstag an des Philip Forney's seyn werden, um alle die Wagen, Gespänn u. Pferde, welche auf die besagten Täge an diesen Plätzen des halben seyn werden zu schätzen; u. welche als denn nicht da seyn, müssen zu Wills's Creek geschätzet werden.

Die Wagen welche zu York u. Forney's geschätzet werden, müssen so gleich nach der schätzung von dort aussetzen nach Wills's Creek zu, unter der Aufsicht und Anordnung der Personen welche ich deswegen darzu ernennen werde.

Der oder die Eigner eines ieglichen Wagens oder Fuhre müssen sich bringen an den Ort der Schätzung, u. den schätzern überliefern, ein schreiben warinnen enthalten, eine Beschreibung ihrer verschiedenen Pferdte, nebst deren verschiedenen natürlichen und gelernten Kennzeichen, welches Verzeichniß dem Contract solle bey gefüget werden.

Jeder Wagen soll versehen seyn miteiner Decke, damit die Güter so darauf geladen werden mögen bewahret bleiben vor dem Regen, und die Gesundtheit der Fuhrleute, welche auf den Wagen logieren müssen, nicht verletzet werde. Und jede Decke soll gezeichnet seyn mit des accordirenden Namen mit grossen Buchstaben, an jedem Wagen: ingleichen soll jeder Pferde Treiber versehen seyn mit einer Sichel, damit das lange Gras welches in dem Lande jenseit der Geburge wächset, kan abgeschnitten werden.

Und da alle Wagen verpflichtet sind eine Latung Haber oder Inschen-Korn zu führen, so werden alle diejenige, so den gleichen zu verkauffen haben, ersucht vorsichtig zu seyn, und nicht durch forderung eines Ubermässigen Preisses desselben, den Dienst des Königs zu verhintern.

B. Fränklin.

Franklin's advertisement about the valuation of wagons for General Braddock's expedition.

from this day to next Wednesday evening, and at York from next Thursday morning till Friday evening, where I shall be ready to agree for wagons and teams, or single horses, on the following terms, viz.: 1. That there shall be paid for each wagon with four good horses and a driver fifteen shillings per diem; and for each able horse with a pack-saddle or other saddle and furniture, two shillings per diem; and for each able horse without a saddle, eighteen pence per diem. 2. That the pay commence from the time of their joining the forces at Will's Creek, which must be on or before the 20th of May ensuing, and that a reasonable allowance be paid over and above for the time necessary for their traveling to Will's Creek and home again after their discharge. 3. Each wagon and team and every saddle or pack horse is to be valued by indifferent [1] persons chosen between me and the owner; and in case of the loss of any wagon, team, or other horse in the service, the price according to such valuation is to be allowed and paid. 4. Seven days' pay is to be advanced and paid in hand by me to the owner of each wagon and team, or horse, at the time of contracting, if required, and the remainder to be paid by General Braddock or by the paymaster of the army at the time of their discharge, or from time to time as it shall be demanded. 5. No drivers of wagons or persons taking care of the hired horses are on any account to be called upon to do the duty of soldiers or be otherwise employed than in conducting or taking

[1] Neutral.

care of their carriages or horses. 6. All oats, Indian corn, or other forage that wagons or horses bring to the camp more than is necessary for the subsistence of the horses is to be taken for the use of the army, and a reasonable price paid for the same.

"Note.—My son William Franklin is empowered to enter into like contracts with any person in Cumberland County. B. FRANKLIN."

"To the Inhabitants of the Counties of Lancaster, York and Cumberland.

"FRIENDS AND COUNTRYMEN,
"Being occasionally at the camp at Frederic a few days since, I found the General and officers extremely exasperated on account of their not being supplied with horses and carriages, which had been expected from this Province as most able to furnish them; but, through the dissensions between our Governor and Assembly, money had not been provided nor any steps taken for that purpose.

"It was proposed to send an armed force immediately into these counties to seize as many of the best carriages and horses as should be wanted and compel as many persons into the service as would be necessary to drive and take care of them.

"I apprehended that the progress of British soldiers through these counties on such an occasion, especially considering the temper they are in and their resentment against us, would be attended with many and great inconveniences to the inhabitants, and therefore more willingly took the trouble of trying first what might be done by fair and equitable means. The people of these back counties have lately complained to the Assembly that a sufficient currency was wanting; you have an opportunity of receiving and dividing among you a very considerable sum; for if the service of this expedition should continue, as it is more than probable it will, for one hundred and twenty days, the hire of these wagons and horses will amount to upward of thirty thousand pounds, which will be paid you in silver and gold of the King's money.

"The service will be light and easy, for the army will scarce march above twelve miles per day, and the wagons and baggage-horses, as they carry those things that are absolutely necessary to the welfare of the army, must march with the army and no faster; and are for the army's sake always placed where they can be most secure, whether in a march or in a camp.

"If you are really, as I believe you are, good and loyal subjects to His Majesty, you may now do a most acceptable service and make it easy to yourselves; for three or four of such as cannot separately spare from the business of their plantations a wagon and four horses and a driver, may do it together, one furnishing the wagon, another one or two horses, and another the driver, and divide the pay proportionably between you; but if you do not this service to your King and country voluntarily, when such good pay and reasonable terms are offered to you, your loyalty will be strongly suspected. The King's business must be done; so many brave troops come so far for your defense must not stand idle through your backwardness to do what may be reasonably expected from you; wagons and horses must be had; violent measures will probably be used, and you will be left to seek for a recompense where you can find it, and your case, perhaps, be little pitied or regarded.

"I have no particular interest in this affair, as, except the satisfaction of endeavoring to do good, I shall have only my labor for my pains. If this method of obtaining the wagons and horses is not likely to succeed, I am obliged to send word to the General in fourteen days; and I suppose Sir John St. Clair, the hussar,[1] with a body of soldiers, will immediately enter the Province for the purpose, which I shall be sorry to hear, because I am very sincerely and truly your friend and well-wisher,

"B. Franklin."

I received of the General about eight hundred pounds, to be disbursed in advance-money to the wagon owners, etc.; but that sum being insufficient, I advanced upward of two hundred pounds more, and in two weeks the one hundred and fifty wagons with two hundred and fifty-nine carry-

[1] Cavalryman.

ing horses were on their march for the camp. The advertisement promised payment according to the valuation in case any wagon or horse should be lost. The owners, however, alleging they did not know General Braddock or what dependence might be had on his promise, insisted on my bond for the performance, which I accordingly gave them.

While I was at the camp supping one evening with the officers of Colonel Dunbar's regiment, he represented to me his concern for the subalterns [1] who, he said, were generally not in affluence and could ill afford in this dear [2] country to lay in the stores that might be necessary in so long a march through a wilderness where nothing was to be purchased. I commiserated their case and resolved to endeavor procuring them some relief. I said nothing, however, to him of my intention, but wrote the next morning to the committee of the Assembly, who had the disposition of some public money, warmly recommending the case of these officers to their consideration and proposing that a present should be sent them of necessaries and refreshments. My son, who had some experience of a camp life, and of its wants, drew up a list for me which I enclosed in my letter. The committee approved and used such diligence that, conducted by my son, the stores arrived at the camp as soon as the wagons. They consisted of twenty parcels, each containing:

[1] Commissioned officer below the rank of captain.
[2] Expensive.

6 lbs. loaf sugar.
6 lbs. good Muscovado ditto.
1 lb. good green tea.
1 lb. good bohea ditto.
6 lbs. good ground coffee.
6 lbs. chocolate.
1-2 cwt. best white biscuit.
1-2 lb. pepper.
1 quart best white wine vinegar.
1 Gloucester cheese.
1 keg containing 20 lbs. good butter.
2 doz. old Madeira wine.
2 gallons Jamaica spirits.
1 bottle flour of mustard.
2 well-cured hams.
1-2 dozen dried tongues.
6 lbs. rice.
6 lbs. raisins.

These twenty parcels, well packed, were placed on as many horses, each parcel with the horse being intended as a present for one officer. They were very thankfully received and the kindness acknowledged by letters to me from the colonels of both regiments in the most graceful terms. The General, too, was highly satisfied with my conduct in procuring him the wagons, etc., and readily paid my account of disbursements, thanking me repeatedly and requesting my farther assistance in sending provisions after him. I undertook this also, and was busily employed in it till we heard of his defeat, advancing for the service of my own money upwards of one thousand pounds sterling, of which I sent him an account. It came to his hands, luckily for me, a few days before the battle, and he returned me immediately an order on the paymaster for the round sum of one thousand pounds, leaving the remainder to the next account. I consider this payment as good luck, having never been able to obtain that remainder, of which more hereafter.

"Too Much Self-Confidence . . ."

THIS General was, I think, a brave man and might probably have made a figure as a good officer in some European war. But he had too much self-confidence, too high an opinion of the validity [1] of regular troops, and too mean a one of both Americans and Indians. George Croghan, our Indian interpreter, joined him on his march with one hundred of those people, who might have been of great use to his army as guides, scouts, etc., if he had treated them kindly; but he slighted and neglected them, and they gradually left him.

In conversation with him one day, he was giving me some account of his intended progress. "After taking Fort Duquesne" [2] says he, "I am to proceed to Niagara; and, having taken that, to Frontenac,[3] if the season will allow time; and I suppose it will, for Duquesne can hardly detain me above three or four days; and then I see nothing that can obstruct my march to Niagara." Having before revolved in my mind the long line his army must make in their march by a very narrow road to be cut for them through the woods and bushes, and also what I had read of a former defeat of fifteen hundred French who invaded the Iroquois country, I had conceived some doubts and some fears for the event of the campaign. But I ventured only to say, "To be sure, sir, if you arrive well before

[1] Worth.   [2] Now Pittsburgh.   [3] Now Kingston, Ontario.

Duquesne with these fine troops, so well provided with artillery, that place, not yet completely fortified and as we hear with no very strong garrison, can probably make but a short resistance. The only danger I apprehend of obstruction to your march is from ambuscades of Indians, who by constant practice are dexterous in laying and executing them; and the slender line, near four miles long, which your army must make may expose it to be attacked by surprise in its flanks and to be cut like a thread into several pieces which from their distance cannot come up in time to support each other."

He smiled at my ignorance, and replied, "These savages may, indeed, be a formidable enemy to your raw American militia, but upon the King's regular and disciplined troops, sir, it is impossible they should make any impression." I was conscious of an impropriety in my disputing with a military man in matters of his profession, and said no more. The enemy, however, did not take the advantage of his army which I apprehended its long line of march exposed it to, but let it advance without interruption till within nine miles of the place; and then, when more in a body (for it had just passed a river, where the front had halted till all were come over), and in a more open part of the woods than any it had passed, attacked its advanced guard by heavy fire from behind trees and bushes, which was the first intelligence the General had of an enemy's being near him. This guard being disordered, the General hurried the troops up to their assistance, which was done in great confusion through wagons, baggage, and cattle; and presently the fire came upon their flank; the officers, being on horseback, were more easily distinguished, picked out as marks, and fell very fast; and the soldiers were crowded together in a huddle, having or hearing no orders and standing to be shot at till two-thirds of them were killed; and then, being seized with a panic, the whole fled with precipitation.

The wagoners took each a horse out of his team and scampered; their example was immediately followed by others, so that all the wagons, provisions, artillery, and stores were left to the enemy. The General, being wounded, was brought off with difficulty; his secretary Mr. Shirley was killed by his side; and out of eighty-six officers, sixty-three were killed or wounded, and seven hundred and fourteen men killed out of eleven hundred. These eleven hundred had been picked men from the whole army; the rest had been left behind with Colonel Dunbar, who was to follow with the heavier part of the stores, provisions, and baggage. The flyers, not being pursued, arrived at Dunbar's camp, and the panic they brought with them instantly seized him and all his people; and though he had now above one thousand men and the enemy who had beaten Braddock did not at most exceed four hundred Indians and French together, instead of proceeding and endeavoring to recover some of the lost honor, he ordered all the stores, ammunition, etc., to be destroyed that he might have more horses to assist his flight towards the settlements and less lumber to remove. He was there met with requests from the governors of Virginia, Maryland, and Pennsylvania that he would post his troops on the frontier so as to afford some protection to the inhabitants; but he continued his hasty march through all the country, not thinking himself safe till he arrived at Philadelphia, where the inhabitants could protect him. This whole transaction gave us Americans the first suspicion that our exalted ideas of the prowess of British regulars had not been well founded.

"Out of Conceit of Such Defenders . . ."

IN THEIR first march, too, from their landing till they got beyond the settlements they had plundered and stripped the inhabitants, totally ruining some poor families, besides insulting, abusing, and confining the people if they remonstrated. This was enough to put us out of conceit [1] of such defenders, if we had really wanted any. How different was the conduct of our French friends in 1781, who, during a march through the most inhabited part of our country from Rhode Island to Virginia, near seven hundred miles, occasioned not the smallest complaint for the loss of a pig, a chicken, or even an apple.

Captain Orme, who was one of the General's aides-de-camp, and, being grievously wounded, was brought off with him and continued with him to his death, which

[1] To make us dissatisfied with.

happened in a few days, told me that he was totally silent all the first day and at night only said, "Who would have thought it?" That he was silent again the following day, saying only at last, "We shall better know how to deal with them another time"; and died in a few minutes after.

The secretary's papers with all the General's orders, instructions, and correspondence, falling into the enemy's hands, they selected and translated into French a number of the articles, which they printed to prove the hostile intentions of the British court before the declaration of war. Among these I saw some letters of the General to the ministry, speaking highly of the great service I had rendered the army and recommending me to their notice. David Hume, too, who was some years after secretary to Lord Hertford when minister in France, and afterward to General Conway when secretary of state, told me he had seen among the papers in that office letters from Braddock highly recommending me. But the expedition having been unfortunate, my service, it seems, was not thought of much value, for those recommendations were never of any use to me.

As to rewards from himself, I asked only one, which was that he would give orders to his officers not to enlist any more of our bought servants,[1] and that he would discharge such as had been already enlisted. This he readily granted, and several were accordingly returned to their masters on my application. Dunbar, when the command

[1] Bound to serve a period of years to pay their passage to America.

devolved on him, was not so generous. He being at Philadelphia on his retreat, or rather flight, I applied to him for the discharge of the servants of three poor farmers of Lancaster County that he had enlisted, reminding him of the late General's orders on that head. He promised me that, if the masters would come to him at Trenton, where he should be in a few days on his march to New York, he would there deliver their men to them. They accordingly were at the expense and trouble of going to Trenton, and there he refused to perform his promise, to their great loss and disappointment.

As soon as the loss of the wagons and horses was generally known, all the owners came upon me for the valuation which I had given bond to pay. Their demands gave me a great deal of trouble; my acquainting them that the money was ready in the paymaster's hands, but that orders for paying it must first be obtained from General Shirley, and my assuring them that I had applied to that General by letter but, he being at a distance, an answer could not soon be received, and they must have patience—all this was not sufficient to satisfy, and some began to sue me. General Shirley at length relieved me from this terrible situation by appointing commissioners to examine the claims and ordering payment. They amounted to near twenty thousand pound, which to pay would have ruined me.

Before we had the news of this defeat, the two Doctors Bond came to me with a subscription paper for raising money to defray the expense of a grand firework which it was intended to exhibit at a rejoicing on receipt of the news of our taking Fort Duquesne. I looked grave and said it would, I thought, be time enough to prepare for the rejoicing when we knew we should have occasion to rejoice. They seemed surprised that I did not immediately comply with their proposal. "Why," says one of them, "you surely don't suppose that the fort will not be taken?" "I don't know that it will not be taken, but I know that the events of war are subject to great uncertainty." I gave them the reasons of my doubting, the subscription was dropped, and the projectors thereby missed the mortification they would have undergone if the firework had been prepared. Dr. Bond on some other occasion afterward said that he did not like Franklin's forebodings.

Governor Morris, who had continually worried the Assembly with message after message before the defeat of Braddock to beat them into the making of acts to raise money for the defense of the Province, without taxing, among others, the proprietary estates, and had rejected all their bills for not having such an exempting clause, now redoubled his attacks with more hope of success, the danger and necessity being greater. The Assembly, however, continued firm, believing they had justice on their side and that it would be giving up an essential right if they suffered the Governor to amend their money-bills. In one of the last, indeed, which was for granting

Autobiography

fifty thousand pounds, his proposed amendment was only of a single word. The bill expressed "that all estates, real and personal, were to be taxed, those of the Proprietaries *not* excepted." His amendment was, for *not* read *only:* a small, but very material alteration. However, when the news of this disaster reached England, our friends there, whom we had taken care to furnish with all the Assembly's answers to the Governor's messages, raised a clamor against the Proprietaries for their meanness and injustice in giving their Governor such instructions, some going so far as to say that by obstructing the defense of their Province they forfeited their right to it. They were intimidated by this and sent orders to their receiver-general to add five thousand pounds of their money to whatever sum might be given by the Assembly for such purpose.

This, being notified to the House, was accepted in lieu of their share of a general tax, and a new bill was formed with an exempting clause, which passed accordingly. By this act I was appointed one of the commissioners for disposing of the money, sixty thousand pounds. I had been active in modeling the bill and procuring its passage and had at the same time drawn a bill for establishing and disciplining a voluntary militia, which I carried through the House without much difficulty, as care was taken in it to leave the Quakers at their liberty. To promote the association necessary to form the militia, I wrote a dialogue stating and answering all the objections I could think of to such a militia, which was printed and had, as I thought, great effect.

"I Undertook Military Business . . ."

WHILE the several companies in the city and country were forming and learning their exercise, the Governor prevailed with me to take charge of our northwestern frontier, which was infested by the enemy, and provide for the defense of the inhabitants by raising troops and building a line of forts. I undertook his military business, though I did not conceive myself well qualified for it. He gave me a commission with full powers and a parcel of blank commissions for officers to be given to whom I thought fit. I had but little difficulty in raising men, having soon five hundred and sixty under my command. My son, who had in the preceding war been an officer in the army raised against Canada, was my aide-de-camp and of great use to me. The Indians had burned Gnadenhut, a village settled by the Moravians, and massacred the inhabitants; but the place was thought a good situation for one of the forts.

In order to march thither, I assembled the companies at Bethlehem, the chief establishment of those people. I was surprised to find it in so good a posture of defense; the destruction of Gnadenhut had made them apprehend danger. The principal buildings were defended by a stockade; they had purchased a quantity of arms and ammunition from New York and had even placed quantities of small paving stones between the windows of their high stone houses for their women to throw down upon the heads of any Indians that should attempt to force into them. The armed brethren, too, kept watch, and relieved [1] as methodically as in any garrison town. In conversation with the bishop, Spangenberg, I mentioned this, my surprise; for, knowing they had obtained an act of Parliament exempting them from military duties in the Colonies, I had supposed they were conscientiously scrupulous of bearing arms. He answered me that it was not one of their established principles, but that at the time of their obtaining that act it was thought to be a principle with many of their people. On this occasion, however, they to their surprise found it adopted by but a few. It seems they were either deceived in themselves or deceived the Parliament; but common sense aided by present danger will sometimes be too strong for whimsical opinions.

It was the beginning of January when we set out upon this business of building forts. I sent one detachment toward the Minisink [2] with instructions to erect one for the security of that upper part of the country, and another to the lower part with similar instructions; and I concluded to go myself with the rest of my force to Gnadenhut, where a fort was thought more immediately necessary. The Moravians procured me five wagons for our tools, stores, baggage, etc.

[1] Changed guard.   [2] Waterfall on Stony Creek.

Just before we left Bethlehem, eleven farmers who had been driven from their plantations by the Indians came to me requesting a supply of firearms that they might go back and fetch off their cattle. I gave them each a gun with suitable ammunition. We had not marched many miles before it began to rain, and it continued raining all day; there were no habitations on the road to shelter us till we arrived near night at the house of a German, where, and in his barn, we were all huddled together, as wet as water could make us. It was well we were not attacked in our march, for our arms were of the most ordinary sort, and our men could not keep their gun locks dry. The Indians are dexterous in contrivances for that purpose, which we had not. They met that day the eleven poor farmers above mentioned and killed ten of them. The one who escaped informed [us] that his and his companions' guns would not go off, the priming [1] being wet with the rain.

The next day being fair, we continued our march and arrived at the desolated Gnadenhut. There was a sawmill near, round which were left several piles of boards, with which we soon hutted ourselves, an operation the more necessary at that inclement season, as we had no tents. Our first work was to bury more effectually the dead we found there who had been half interred by the country people.

[1] The flintlocks were primed with powder ignited by a spark from t flint.

Autobiography

The next morning our fort was planned and marked out, the circumference measuring four hundred and fifty-five feet, which would require as many palisades to be made of trees one with another of a foot diameter each. Our axes, of which we had seventy, were immediately set to work to cut down trees and, our men being dexterous in the use of them, great dispatch was made. Seeing the trees fall so fast, I had the curiosity to look at my watch when two men began to cut at a pine; in six minutes they had it upon the ground, and I found it of fourteen inches diameter. Each pine made three palisades of eighteen feet long, pointed at one end. While these were preparing, our other men dug a trench all round of three feet deep in which the palisades were to be planted; and, our wagons, the bodies being taken off and the fore and hind wheels separated by taking out the pin which united the two parts of the perch,[1] we had ten carriages with two horses each to bring the palisades from the woods to the spot. When they were set up, our carpenters built a stage of boards all round within about six feet high for the men to stand on when to fire through the loopholes. We had one swivel gun,[2] which we mounted on one of the angles and fired it as soon as fixed to let the Indians know, if any were within hearing, that we had such pieces; and thus our fort, if such a magnificent name may be given to so miserable

[1] A pole connecting the fore and hind axles of a wagon.
[2] A small cannon that could be turned on a swivel in different directions.

a stockade, was finished in a week, though it rained so hard every other day that the men could not work.

This gave me occasion to observe that when men are employed they are best contented; for on the days they worked they were good-natured and cheerful, and with the consciousness of having done a good day's work they spent the evening jollily; but on our idle days they were mutinous and quarrelsome, finding fault with their pork, the bread, etc., and in continual ill-humor, which put me in mind of a sea captain, whose rule it was to keep his men constantly at work; and when his mate once told him that they had done everything and there was nothing further to employ them about, "Oh," says he, "make them scour the anchor."

This kind of fort, however contemptible, is a sufficient defense against Indians who have no cannon. Finding ourselves now posted securely and having a place to retreat to on occasion, we ventured out in parties to scour the adjacent country. We met with no Indians, but we found the places on the neighboring hills where they had lain to watch our proceedings. There was an art in their contrivance of those places that seems worth mention. It being winter, a fire was necessary for them; but a common fire on the surface of the ground would by its light have discovered their position at a distance. They had therefore dug holes in the ground about three feet diameter and somewhat deeper; we saw where they had with their hatchets cut off the charcoal from the sides of burnt logs lying in the woods. With these coals they had made small fires in the bottom of the holes, and we observed among the weeds and grass the prints of their bodies made by their lying all round with their legs hanging down in the holes to keep their feet warm, which with them is an essential point. This kind of fire, so managed, could not discover them, either by its light, flame, sparks, or even smoke: it appeared that their number was not great, and it seems they saw we were too many to be attacked by them with prospect of advantage.

We had for our chaplain a zealous Presbyterian minister, Mr. Beatty, who complained to me that the men did not generally attend his prayers and exhortations. When they enlisted they were promised besides pay and provisions a gill of rum a day, which was punctually served out to them, half in the morning and the other half in the evening; and I observed they were as punctual in attending to receive it; upon which I said to Mr. Beatty, "It is, perhaps, below the dignity of your profession to act as steward of the rum, but if you were to deal it out and only just after prayers, you would have them all about you." He liked the thought, undertook the office, and with the help of a few hands to measure out the liquor executed it to satisfaction, and never were prayers more generally and more punctually attended; so that I thought this method preferable to the punishment inflicted by some military laws for non-attendance on divine service.

I had hardly finished this business and got my fort well

stored with provisions, when I received a letter from the Governor acquainting me that he had called the Assembly and wished my attendance there, if the posture of affairs on the frontiers was such that my remaining there was no longer necessary. My friends, too, of the Assembly, pressing me by their letters to be if possible at the meeting, and my three intended forts being now completed, and the inhabitants contented to remain on their farms under that protection, I resolved to return; the more willingly as a New England officer, Colonel Clapham, experienced in Indian war, being on a visit to our establishment, consented to accept the command. I gave him a commission and, parading the garrison, had it read before them and introduced him to them as an officer who from his skill in military affairs was much more fit to command them than myself, and, giving them a little exhortation, took my leave. I was escorted as far as Bethlehem, where I rested a few days to recover from the fatigue I had undergone. The first night, being in a good bed, I could hardly sleep, it was so different from my hard lodging on the floor of our hut at Gnaden wrapped only in a blanket or two.

"*The Practice of the Moravians . . .*"

WHILE at Bethlehem, I inquired a little into the practice of the Moravians; some of them had accompanied me, and all were very kind to me. I found they worked for a common stock, ate at common tables, and slept in common

Autobiography

dormitories, great numbers together. In the dormitories I observed loopholes at certain distances all along just under the ceiling, which I thought judiciously placed for change of air. I was at their church, where I was entertained with good music, the organ being accompanied with violins, hautboys,[1] flutes, clarinets, etc. I understood that their sermons were not usually preached to mixed congregations of men, women, and children as is our common practice, but that they assembled sometimes the married men, at other times their wives, then the young men, the young women, and the little children, each division by itself. The sermon I heard was to the latter, who came in and were placed in rows on benches; the boys under the conduct of a young man, their tutor, and the girls conducted by a young woman. The discourse seemed well adapted to their capacities, and was delivered in a pleasing, familiar manner, coaxing them, as it were, to be good. They behaved very orderly, but looked pale and unhealthy, which made me suspect they were kept too much within doors or not allowed sufficient exercise.

I inquired concerning the Moravian marriages, whether the report was true that they were by lot. I was told that lots were used only in particular cases; that generally when a young man found himself disposed to marry, he informed the elders of his class, who consulted the elder ladies that governed the young women. As these elders of the different sexes were well acquainted with the tempers

[1] Oboes.

and dispositions of their respective pupils, they could best judge what matches were suitable, and their judgments were generally acquiesced in; but if, for example, it should happen that two or three young women were found to be equally proper for the young man, the lot was then recurred to. I objected, if the matches are not made by mutual choice of the parties, some of them may chance to be very unhappy. "And so they may," answered my informer, "if you let the parties choose for themselves"; which, indeed, I could not deny.

Being returned to Philadelphia, I found the association went on swimmingly, the inhabitants that were not Quakers having pretty generally come into it, formed themselves into companies, and chose their captains, lieutenants, and ensigns [1] according to the new law. Dr. B. visited me, and gave me an account of the pains he had taken to spread a general good liking to the law and ascribed much to those endeavors. I had had the vanity to ascribe all to my *Dialogue*;[2] however, not knowing but that he might be in the right, I let him enjoy his opinion, which I take to be generally the best way in such cases. The officers, meeting, chose me to be colonel of the regiment, which I this time accepted. I forget how many companies we had, but we paraded about twelve hundred well-looking men with a company of artillery, who had been furnished with

[1] Standard bearers.
[2] Franklin's plan for a militia appeared in his *Gazette*, in the form of *A Dialogue between X, Y, and Z concerning the Present State of Affairs in Pennsylvania.*

six brass field pieces,[1] which they had become so expert in the use of as to fire twelve times in a minute. The first time I reviewed my regiment they accompanied me to my house and would salute me with some rounds fired before my door, which shook down and broke several glasses of my electrical apparatus. And my new honor proved not much less brittle; for all our commissions were soon after broken by a repeal of the law in England.

During this short time of my colonelship, being about to set out on a journey to Virginia, the officers of my regiment took it into their heads that it would be proper for them to escort me out of town as far as the Lower Ferry. Just as I was getting on horseback they came to my door, between thirty and forty, mounted and all in their uniforms. I had not been previously acquainted with the project or I should have prevented it, being naturally averse to the assuming of state on any occasion; and I was a good deal chagrined at their appearance, as I could not avoid their accompanying me. What made it worse was that as soon as we began to move they drew their swords and rode with them naked all the way. Somebody wrote an account of this to the Proprietor, and it gave him great offense. No such honor had been paid him when in the Province, nor to any of his governors; and he said it was only proper to princes of the blood royal, which may be true for aught I know, who was and still am ignorant of the etiquette in such cases.

[1] Cannon upon wheels.

This silly affair, however, greatly increased his rancor against me, which was before not a little on account of my conduct in the Assembly respecting the exemption of his estate from taxation, which I had always opposed very warmly and not without severe reflections on his meanness and injustice of contending for it. He accused me to the ministry as being the great obstacle to the King's service, preventing by my influence in the House the proper form of the bills for raising money, and he instanced this parade with my officers as a proof of my having an intention to take the government of the Province out of his hands by force. He also applied to Sir Everard Fawkener, the postmaster-general, to deprive me of my office; but it had no other effect than to procure from Sir Everard a gentle admonition.

Notwithstanding the continual wrangle between the Governor and the House in which I as a member had so large a share, there still subsisted a civil intercourse between that gentleman and myself, and we never had any personal difference. I have sometimes since thought that his little or no resentment against me for the answers it was known I drew up to his messages might be the effect of professional habit, and that, being bred a lawyer, he might consider us both as merely advocates for contending clients in a suit, he for the Proprietaries and I for the Assembly. He would, therefore, sometimes call in a friendly way to advise with me on difficult points, and sometimes, though not often, take my advice.

We acted in concert to supply Braddock's army with provisions, and when the shocking news arrived of his defeat, the Governor sent in haste for me to consult with him on measures for preventing the desertion of the back counties. I forget now the advice I gave; but I think it was that Dunbar should be written to and prevailed with if possible to post his troops on the frontiers for their protection, till by reinforcements from the colonies he might be able to proceed on the expedition. And after my return from the frontier, he would have had me undertake the conduct of such an expedition with provincial troops for the reduction of Fort Duquesne, Dunbar and his men being otherwise employed; and he proposed to commission me as general. I had not so good an opinion of my military abilities as he professed to have, and I believe his professions must have exceeded his real sentiments; but probably he might think that my popularity would facilitate the raising of the men and my influence in Assembly, the grant of money to pay them, and that perhaps without taxing the proprietary estate. Finding me not so forward to engage as he expected, the project was dropped, and he soon after left the government, being superseded by Captain Denny.

"Drawing Lightning from the Clouds . . ."

BEFORE I proceed in relating the part I had in public affairs under this new Governor's administration, it may not be amiss here to give some account of the rise and progress of my philosophical reputation.

In 1746, being at Boston, I met there with a Dr. Spence, who was lately arrived from Scotland and showed me some electric experiments. They were imperfectly performed, as he was not very expert; but, being on a subject quite new to me, they equally surprised and pleased me. Soon after my return to Philadelphia our library company received from Mr. P. Collinson, Fellow of the Royal Society of London, a present of a glass tube with some account of the use of it in making such experiments. I eagerly seized the opportunity of repeating what I had seen at Boston; and by much practice acquired great readiness in performing those also which we had an account of from England, adding a number of new ones. I say much practice, for my house was continually full for some time with people who came to see these new wonders.

To divide a little this encumbrance among my friends, I caused a number of similar tubes to be blown at our glass-house, with which they furnished themselves, so that we had at length several performers. Among these the principal was Mr. Kinnersley, an ingenious neighbor, who, being out of business, I encouraged to undertake showing the experiments for money, and drew up for him two lectures in which the experiments were ranged in such order and accompanied with such explanations in such method as that the foregoing should assist in comprehending the following. He procured an elegant apparatus for the purpose in which all the little machines that I had roughly made for myself were nicely formed by instrument-makers. His lectures were well attended and gave great satisfaction; and after some time he went through the colonies exhibiting them in every capital town and picked up some money. In the West India Islands indeed it was with difficulty the experiments could be made, from the general moisture of the air.

Obliged as we were to Mr. Collinson for his present of the tube, etc., I thought it right he should be informed of our success in using it and wrote him several letters containing accounts of our experiments. He got them read in the Royal Society, where they were not at first thought worth so much notice as to be printed in their Transactions. One paper which I wrote for Mr. Kinnersley on the sameness of lightning with electricity I sent to Dr. Mitchel, an acquaintance of mine, and one of the members also of that society, who wrote me word that it had been read, but was laughed at by the connoisseurs. The papers, however, being shown to Dr. Fothergill, he thought them of too much value to be stifled and advised the printing of them. Mr. Collinson then gave them to Cave for publication in his *Gentleman's Magazine;* but he chose to print them separately in a pamphlet, and Dr. Fothergill wrote the preface. Cave, it seems, judged rightly for his profit, for by the additions that arrived afterward they swelled to a quarto volume, which has had five editions and cost him nothing for copy-money.[1]

It was, however, some time before those papers were much taken notice of in England. A copy of them happening to fall into the hands of the Count de Buffon, a philosopher deservedly of great reputation in France and indeed all over Europe, he prevailed with M. Dalibard to translate them into French, and they were printed at Paris. The publication offended the Abbé Nollet, preceptor in natural philosophy to the royal family, and an able experimenter, who had formed and published a theory of electricity which then had the general vogue. He could not at first believe that such a work came from America, and said it must have been fabricated by his enemies at Paris to decry his system. Afterwards, having been assured that there really existed such a person as Franklin at Philadelphia, which he had doubted, he wrote and published a volume of letters chiefly addressed to me, defending his theory and denying the verity of my experiments and of the positions deduced from them.

I once purposed answering the Abbé and actually began the answer; but, on consideration that my writings contained a description of experiments which any one might repeat and verify, and if not to be verified, could not be defended; or of observations offered as conjectures and not delivered dogmatically, therefore not laying me under any obligation to defend them; and reflecting that a dis-

[1] Royalty paid to the author.

pute between two persons writing in different languages might be lengthened greatly by mistranslations, and thence misconceptions of one another's meaning, much of one of the Abbé's letters being founded on an error in the translation, I concluded to let my papers shift for themselves, believing it was better to spend what time I could spare from public business in making new experiments, than in disputing about those already made. I therefore never answered M. Nollet, and the event gave me no cause to repent my silence; for my friend M. le Roy of the Royal Academy of Sciences took up my cause and refuted him; my book was translated into the Italian, German and Latin languages; and the doctrine it contained was by degrees universally adopted by the philosophers of Europe in preference to that of the Abbé, so that he lived to see himself the last of his sect, except Monsieur B—— of Paris, his *élève* and immediate disciple.

What gave my book the more sudden and general celebrity was the success of one of its proposed experiments, made by Messrs. Dalibard and De Lor at Marly for drawing lightning from the clouds. This engaged the public attention everywhere. M. de Lor, who had an apparatus for experimental philosophy and lectured in that branch of science, undertook to repeat what he called the *Philadelphia Experiments;* and, after they were performed before the King and court, all the curious of Paris flocked to see them. I will not swell this narrative with an account of that capital experiment, nor of the infinite pleasure I received in the success of a similar one I made soon after with a kite at Philadelphia, as both are to be found in the histories of electricity.

Dr. Wright, an English physician, when at Paris, wrote to a friend, who was of the Royal Society, an account of the high esteem my experiments were in among the learned abroad, and of their wonder that my writings had been so little noticed in England. The Society, on this, resumed the consideration of the letters that had been read to them; and the celebrated Dr. Watson drew up a summary account of them and of all I had afterwards sent to England on the subject, which he accompanied with some praise of the writer. This summary was then printed in their Transactions; and some members of the Society in London, particularly the very ingenious Mr. Canton, having verified the experiment of procuring lightning from the clouds by a pointed rod and acquainting them with the success, they soon made me more than amends for the slight with which they had before treated me. Without my having made any application for that honor they chose me a member and voted that I should be excused the customary payments, which would have amounted to twenty-five guineas, and ever since have given me their Transactions *gratis*. They also presented me with the gold medal of Sir Godfrey Copley for the year 1753, the delivery of which was accompanied by a very handsome speech of the president, Lord Macclesfield, wherein I was highly honored.

Statesman
1756-1790

Statesman

OUR NEW Governor, Captain Denny, brought over for me the before-mentioned medal from the Royal Society, which he presented to me at an entertainment given him by the city. He accompanied it with very polite expressions of his esteem for me, having, as he said, been long acquainted with my character. After dinner, when the company, as was customary at that time, were engaged in drinking, he took me aside into another room and acquainted me that he had been advised by his friends in England to cultivate a friendship with me as one who was capable of giving him the best advice and of contributing most effectually to the making his administration easy; that he therefore desired of all things to have a good understanding with me, and he begged me to be assured of his readiness on all occasions to render me every service that might be in his power. He said much to me, also, of the Proprietor's good disposition towards the Province and of the advantage it might be to us all and to me in particular if the opposition that had been so long continued to his measures was dropped and harmony restored between him and the people; in effecting which, it was thought no one could be more serviceable than myself; and I might depend on adequate acknowledgments and recompenses, etc., etc. The drinkers, finding we did not return immediately to the table, sent us a decanter of madeira, which the Governor made liberal use of and in proportion became more profuse of his solicitations and promises.

My answers were to this purpose: that my circumstances, thanks to God, were such as to make proprietary favors unnecessary to me; and that being a member of the Assembly I could not possibly accept of any; that, however, I had no personal enmity to the Proprietary, and that whenever the public measures he proposed should appear to be for the good of the people, no one should espouse and forward them more zealously than myself; my past opposition having been founded on this, that the measures which had been urged were evidently intended to serve the proprietary interest, with great prejudice to that of the people; that I was much obliged to him (the Governor) for his professions of regard to me and that he might rely on everything in my power to make his administration as easy as possible, hoping at the same time that he had not brought with him the same unfortunate instructions his predecessor had been hampered with.

On this he did not then explain himself; but when he afterwards came to do business with the Assembly, they appeared again, the disputes were renewed, and I was as active as ever in the opposition, being the penman, first of the request to have a communication of the instructions and then of the remarks upon them, which may be found in the votes of the time, and in the Historical Review I afterward published. But between us personally no enmity arose; we were often together; he was a man of letters, had seen much of the world, and was very entertaining and pleasing in conversation. He gave me the first information that my old friend James Ralph was still alive; that he was esteemed one of the best political writers in England; had been employed in the dispute between Prince Frederick and the King, and had obtained a pension of three hundred a year; that his reputation was indeed small as a poet, Pope having damned his poetry in the *Dunciad;* [1] but his prose was thought as good as any man's.

The Assembly finally, finding the Proprietary obstinately persisted in manacling their deputies [2] with instructions inconsistent not only with the privileges of the people, but with the service of the Crown, resolved to petition the King against them and appointed me their agent to go over to England to present and support the

[1] Pope alludes to Ralph's poem *Night*, 1728.
Silence, ye wolves, while Ralph to Cynthia howls,
And makes Night hideous—Answer him, ye owls!
iii, 165-166.

[2] Governors.

petition. The House had sent up a bill to the Governor granting a sum of sixty thousand pounds for the King's use (ten thousand pounds of which was subjected to the orders of the then general, Lord Loudoun) which the Governor absolutely refused to pass, in compliance with his instructions.

I had agreed with Captain Morris of the packet at New York for my passage, and my stores were put on board, when Lord Loudoun arrived at Philadelphia, expressly, as he told me, to endeavor an accommodation between the Governor and Assembly, that His Majesty's service might not be obstructed by their dissensions. Accordingly, he desired the Governor and myself to meet him that he might hear what was to be said on both sides. We met and discussed the business. In behalf of the Assembly, I urged all the various arguments that may be found in the public papers of that time which were of my writing and are printed with the minutes of the Assembly, and the Governor pleaded his instructions, the bond he had given to observe them, and his ruin if he disobeyed, yet seemed not unwilling to hazard himself if Lord Loudoun would advise it. This his lordship did not choose to do, though I once thought I had nearly prevailed with him to do it; but finally he rather chose to urge the compliance of the Assembly; and he entreated me to use my endeavors with them for that purpose, declaring that he would spare none of the King's troops for the defense of our frontiers and that if we did not continue to provide for that defense ourselves, they must remain exposed to the enemy.

I acquainted the House with what had passed and, presenting them with a set of resolutions I had drawn up declaring our rights and that we did not relinquish our claim to those rights, but only suspended the exercise of them on this occasion through *force*, against which we protested, they at length agreed to drop that bill and frame another comfortable to the proprietary instructions. This, of course, the Governor passed, and I was then at liberty to proceed on my voyage. But in the meantime the packet had sailed with my sea-stores, which was some loss to me, and my only recompense was his lordship's thanks for my service, all the credit of obtaining the accommodation falling to his share.

"Daily Expectation of Sailing . . ."

HE SET OUT for New York before me; and as the time for dispatching the packet-boats was at his disposition and there were two then remaining there, one of which, he said, was to sail very soon, I requested to know the precise time, that I might not miss her by any delay of mine. His answer was, "I have given out that she is to sail on Saturday next; but I may let you know, *entre nous*, that if you are there by Monday morning you will be in time, but do not delay longer." By some accidental hindrance at a ferry, it was Monday noon before I arrived, and I was much afraid she might have sailed, as the wind was fair; but I was soon made easy by the information that she was still in the harbor and would not move till the next day. One would imagine that I was now on the very point of departing for Europe. I thought so; but I was not then so well acquainted with his lordship's character, of which *indecision* was one of the strongest features. I shall give some instances. It was about the beginning of April that I came to New York, and I think it was near the end of June before we sailed. There were then two of the packet-boats, which had been long in port but were detained for the General's letters, which were always to be ready tomorrow. Another packet arrived; she too was detained; and before we sailed a fourth was expected. Ours was the first to be dispatched, as having been there longest. Passengers were engaged in all, and some extremely impatient to be gone, and the merchants uneasy about their letters, and the orders they had given for insurance (it being war time) for fall goods; but their anxiety availed nothing; his lordship's letters were not ready; and yet whoever waited on him found him always at his desk, pen in hand, and concluded he must needs write abundantly.

Going myself one morning to pay my respects, I found in his antechamber one Innis, a messenger of Philadelphia, who had come from thence express with a packet from Governor Denny for the General. He delivered to me some letters from my friends there, which occasioned my inquiring when he was to return and where he lodged, that I might send some letters by him. He told me he was ordered to call tomorrow at nine for the General's answer to the Governor and should set off immediately. I put my letters into his hands the same day. A fortnight after I met him again in the same place. "So, you are soon returned, Innis?" "*Returned!* no, I am not *gone* yet." "How so?" "I have called here by order every morning these two weeks past for his lordship's letter, and it is not yet ready." "Is it possible, when he is so great a writer? for I see him constantly at his escritoire."[1] "Yes," says Innis, "but he is like St. George on the signs, *always on horseback, and never rides on.*" This observation of the messenger was, it seems, well founded; for when in England I understood that Mr. Pitt gave it as one reason for removing this General and sending Generals Amherst and Wolfe *that the minister never heard from him and could not know what he was doing.*

This daily expectation of sailing and all the three packets going down to Sandy Hook to join the fleet there, the passengers thought it best to be on board, lest by a sudden order the ships should sail, and they be left behind. There, if I remember right, we were about six weeks, consuming our sea-stores, and obliged to procure more. At length the fleet sailed, the General and all his army on board, bound to Louisburg,[2] with the intent to besiege and take that

[1] Writing desk.
[2] Louisburg, on Cape Breton, Nova Scotia, was fortified by the French after 1713, taken by the Massachusetts men in 1745, restored to the French in 1748, and taken again by the English in 1758.

fortress; all the packet-boats in company ordered to attend the General's ship, ready to receive his dispatches when they should be ready. We were out five days before we got a letter with leave to part, and then our ship quitted the fleet and steered for England. The other two packets he still detained, carried them with him to Halifax, where he stayed some time to exercise the men in sham attacks upon sham forts, then altered his mind as to besieging Louisburg and returned to New York with all his troops, together with the two packets above mentioned and all their passengers! During his absence the French and savages had taken Fort George, on the frontier of that Province, and the savages had massacred many of the garrison after capitulation.

I saw afterwards in London Captain Bonnell, who commanded one of those packets. He told me that when he had been detained a month he acquainted his lordship that his ship was grown foul to a degree that must necessarily hinder her fast sailing, a point of consequence for a packet-boat, and requested an allowance of time to heave her down and clean her bottom. He was asked how long time that would require. He answered, three days. The General replied, "If you can do it in one day, I give leave; otherwise not; for you must certainly sail the day after tomorrow." So he never obtained leave, though detained afterwards from day to day during full three months.

I saw also in London one of Bonnell's passengers, who was so enraged against his lordship for deceiving and detaining him so long at New York and then carrying him to Halifax and back again that he swore he would sue him for damages. Whether he did or not, I never heard; but as he represented the injury to his affairs it was very considerable.

On the whole, I wondered much how such a man came to be intrusted with so important a business as the conduct of a great army; but, having since seen more of the great world and the means of obtaining and motives for giving places, my wonder is diminished. General Shirley, on whom the command of the army devolved upon the death of Braddock, would, in my opinion, if continued in place, have made a much better campaign than that of Loudoun in 1757, which was frivolous, expensive, and disgraceful to our nation beyond conception; for, though Shirley was not a bred soldier, he was sensible and sagacious in himself, and attentive to good advice from others, capable of forming judicious plans, and quick and active in carrying them into execution. Loudoun, instead of defending the colonies with his great army, left them totally exposed while he paraded idly at Halifax, by which means Fort George was lost; besides, he deranged all our mercantile operations and distressed our trade by a long embargo on the exportation of provisions on pretense of keeping supplies from being obtained by the enemy, but in reality for beating down their price in favor of the contractors, in whose profits, it was said, perhaps from suspicion only, he had a share. And, when at length the embargo was taken off, by neglecting to send notice of it to Charlestown, the Carolina fleet was detained near three months longer, whereby their bottoms were so much damaged by the worm that a great part of them foundered in their passage home.

Shirley was, I believe, sincerely glad of being relieved from so burdensome a charge as the conduct of an army must be to a man unacquainted with military business. I was at the entertainment given by the city of New York to Lord Loudoun on his taking upon him the command. Shirley, though thereby superseded, was present also. There was a great company of officers, citizens, and strangers, and, some chairs having been borrowed in the neighborhood, there was one among them very low, which fell to the lot of Mr. Shirley. Perceiving it as I sat by him, I said, "They have given you, sir, too low a seat." "No matter," says he, "Mr. Franklin, I find *a low seat* the easiest."

While I was, as afore mentioned, detained at New York, I received all the accounts of the provisions, etc., that I had furnished to Braddock, some of which accounts could not sooner be obtained from the different persons I had employed to assist in the business. I presented them to Lord Loudoun, desiring to be paid the balance. He caused them to be regularly examined by the proper officer, who, after comparing every article with its voucher, certified them to be right and the balance due, for which his lordship promised to give me an order on the paymaster. This was, however, put off from time to time; and, though I called often for it by appointment, I did not get it. At length, just before my departure he told me he had on better consideration concluded not to mix his accounts with those of his predecessors. "And you," says he, "when in England, have only to exhibit your accounts at the treasury and you will be paid immediately."

I mentioned, but without effect, the great and unexpected expense I had been put to by being detained so long at New York as a reason for my desiring to be presently paid; and on my observing that it was not right I should be put to any further trouble or delay in obtaining the money I had advanced, as I charged no commission for my service. "O, Sir," says he, "you must not think of persuading us that you are no gainer; we understand better those affairs, and know that every one concerned in supplying the army finds means in the doing it to fill his own pockets." I assured him that was not my case, and that I had not pocketed a farthing; but he appeared clearly not to believe me; and, indeed, I have since learned that immense fortunes are often made in such employments. As to my balance, I am not paid it to this day, of which more hereafter.

"*View of a Vacant Ocean . . .*"

OUR CAPTAIN of the packet had boasted much, before we sailed, of the swiftness of his ship; unfortunately, when we came to sea, she proved the dullest of ninety-six sail, to his no small mortification. After many conjectures respecting the cause, when we were near another ship almost as dull as ours, which, however, gained upon us, the captain ordered all hands to come aft, and stand as near the ensign staff as possible. We were, passengers included, about forty persons. While we stood there, the ship mended her pace, and soon left her neighbor far behind, which proved clearly what our captain suspected, that she was loaded too much by the head. The casks of water, it seems, had been all placed forward; these he therefore ordered to be moved further aft, on which the ship recovered her character, and proved the best sailer in the fleet.

The captain said she had once gone at the rate of thirteen knots, which is accounted thirteen miles per hour. We had on board as a passenger Captain Kennedy of the Navy, who contended that it was impossible, and that no ship ever sailed so fast, and that there must have been some error in the division of the log-line, or some mistake in heaving the log. A wager ensued between the two captains, to be decided when there should be sufficient wind. Kennedy thereupon examined rigorously the log-line, and, being satisfied with that, he determined to throw the log himself. Accordingly, some days after, when the wind blew very fair and fresh, and the captain of the packet, Lutwidge, said he believed she then went at the rate of thirteen knots, Kennedy made the experiment and owned his wager lost.

The above fact I give for the sake of the following observation. It has been remarked as an imperfection in the art of ship-building that it can never be known, till she is tried, whether a new ship will or will not be a good sailer; for that the model of a good-sailing ship has been exactly followed in a new one which has proved, on the contrary, remarkably dull. I apprehend that this may partly be occasioned by the different opinions of seamen respecting the modes of lading, rigging, and sailing of a ship; each has his system; and the same vessel laden by the judgment and orders of one captain shall sail better or worse than when by the orders of another. Besides, it scarce ever happens that a ship is formed, fitted for the sea, and sailed by the same person. One man builds the hull, another rigs her, a third lades and sails her. No one of these has the advantage of knowing all the ideas and experience of the others, and, therefore, cannot draw just conclusions from a combination of the whole.

Even in the simple operation of sailing when at sea, I have often observed different judgments in the officers who commanded the successive watches, the wind being the same. One would have the sails trimmed sharper or flatter than another, so that they seemed to have no certain rule to govern by. Yet I think a set of experiments might be instituted, first, to determine the most proper form of the hull for swift sailing; next, the best dimensions and properest place for the masts; then the form and quantity of sails, and their position, as the wind may be; and, lastly, the disposition of the lading. This is an age of experiments, and I think a set accurately made and combined would be of great use. I am persuaded, therefore, that ere long some ingenious philosopher will undertake it, to whom I wish success.

We were several times chased [1] in our passage, but outsailed everything, and in thirty days had soundings. We had a good observation, and the captain judged himself so near our port, Falmouth, that, if we made a good run in the night, we might be off the mouth of that harbor in the morning, and by running in the night might escape the notice of the enemy's privateers, who often cruised near the entrance of the channel. Accordingly, all the sail was set that we could possibly make, and the wind being very fresh and fair, we went right before it, and made great way. The captain, after his observation, shaped his course, as he thought, so as to pass wide of the Scilly Isles; but it seems there is sometimes a strong indraught setting up St. George's Channel, which deceives seamen and caused the loss of Sir Cloudesley Shovel's squadron.[2] This indraught was probably the cause of what happened to us.

We had a watchman placed in the bow to whom they often called, "Look well out before there," and he as often answered, "Ay, ay"; but perhaps had his eyes shut and

[1] By French warships. [2] In 1707.

was half asleep at the time, they sometimes answering, as is said, mechanically; for he did not see a light just before us which had been hid by the studding-sails from the man at the helm and from the rest of the watch, but by an accidental yaw of the ship was discovered, and occasioned a great alarm, we being very near it, the light appearing to me as big as a cart wheel. It was midnight and our captain fast asleep; but Captain Kennedy, jumping upon deck, and seeing the danger, ordered the ship to wear round, all sails standing; an operation dangerous to the masts, but it carried us clear and we escaped shipwreck, for we were running right upon the rocks on which the lighthouse was erected. This deliverance impressed me strongly with the utility of lighthouses and made me resolve to encourage the building more of them in America if I should live to return there.

In the morning it was found by the soundings, etc., that we were near our port, but a thick fog hid the land from our sight. About nine o'clock the fog began to rise and seemed to be lifted up from the water like the curtain at a play-house, discovering underneath the town of Falmouth, the vessels in its harbor, and the fields that surrounded it. This was a most pleasing spectacle to those who had been so long without any other prospects than the uniform view of a vacant ocean, and it gave us the more pleasure as we were now free from the anxieties which the state of war occasioned.

"We Arrived in London . . ."

I SET OUT immediately with my son for London, and we only stopped a little by the way to view Stonehenge on Salisbury Plain and Lord Pembroke's house and gardens, with his very curious antiquities at Wilton. We arrived in London the 27th of July, 1757.

[*Here ends the* Autobiography, *as published by William Temple Franklin and his successors. What follows was written in the last year of Franklin's life and was first printed in Bigelow's edition in 1868.*]

As soon as I was settled in a lodging Mr. Charles had provided for me, I went to visit Dr. Fothergill, to whom I was strongly recommended and whose counsel respecting my proceedings I was advised to obtain. He was against an immediate complaint to government and thought the Proprietaries should first be personally applied to, who might possibly be induced by the interposition and persuasion of some private friends to accommodate matters amicably. I then waited on my old friend and correspondent, Mr. Peter Collinson, who told me that John Hanbury, the great Virginia merchant, had requested to be informed when I should arrive that he might carry me to Lord Granville's, who was then President of the Council and wished to see me as soon as possible. I agreed to go with him the next morning. Accordingly Mr. Hanbury called for me and took me in his carriage to that nobleman's, who received me with great civility; and after some questions respecting the present state of affairs in America and discourse thereupon, he said to me: "You Americans have wrong ideas of the nature of your constitution; you contend that the King's instructions to his Governors are not laws, and think yourselves at liberty to regard or disregard them at your own discretion. But those instructions are not like the pocket instructions given to a minister going abroad, for regulating his conduct in some trifling point of ceremony. They are first drawn up by judges learned in the laws; they are then considered, debated, and perhaps amended in Council, after which they are signed by the King. They are then, so far as they relate to you, the *law of the land*, for the King is the Legislator of the Colonies." I told his lordship this was new doctrine to me. I had always understood from our charters that our laws were to be made by our Assemblies, to be presented indeed to the King for his royal assent, but, that being once given, the King could not repeal or alter them. And as the Assemblies could not make permanent laws without his assent, so neither could he make a law for them without theirs. He assured me I was totally mistaken. I did not think so, however, and his lordship's conversation having a little alarmed me as to what might be the sentiments of the court concerning us, I wrote it down as soon as I returned to my lodgings. I recollected that about twenty years before a clause in a bill brought into Parliament by the ministry had proposed to make the King's instructions laws in the colonies, but the clause was thrown out by the Commons, for which we adored them as our friends and friends of liberty, till by their conduct towards us in 1765 it seemed that they had refused that point of sovereignty to the King only that they might reserve it for themselves.

After some days, Dr. Fothergill having spoken to the Proprietaries, they agreed to a meeting with me at Mr. T. Penn's house in Spring Garden. The conversation at first consisted of mutual declarations of disposition to reasonable accommodations, but I suppose each party had its own ideas of what should be meant by *reasonable*. We then went into consideration of our several points of complaint, which I enumerated. The Proprietaries justified their conduct as well as they could, and I the Assembly's. We now appeared very wide, and so far from each other in our opinions as to discourage all hope of agreement. However, it was concluded that I should give them the heads of our complaints in writing, and they promised then to consider them. I did so soon after, but they put the paper into the hands of their solicitor, Ferdinand John Paris, who managed for them all their law business in their great suit with the neighboring proprietary of Maryland, Lord Baltimore, which had subsisted seventy years, and wrote for them all their papers and messages in their dispute with the Assembly. He was a proud, angry man, and as I had occasionally in the answers of the Assembly treated his papers with some severity, they being really weak in point of argument and haughty in expression, he had conceived a mortal enmity to me, which discovering itself whenever we met, I declined the Proprietary's proposal that he and I should discuss the heads of complaint between our two selves, and refused treating with any one but them. They then by his advice put the paper into the hands of the attorney and solicitor-general for their opinion and counsel upon it, where it lay unanswered a year wanting eight days, during which time I made frequent demands of an answer from the Proprietaries, but without obtaining any other than that they had not yet received the opinion of the attorney and solicitor-general. What it was when they did receive it I never learned, for they did not communicate it to me, but sent a long message to the Assembly drawn and signed by Paris, reciting my paper, complaining of its want of formality as a rudeness on my part, and giving a flimsy justification of their conduct, adding that they should be willing to accommodate matters if the Assembly would send out *some person of candor* to treat with them for that purpose, intimating thereby that I was not such.

The want of formality or rudeness was, probably, my not having addressed the paper to them with their assumed titles of True and Absolute Proprietaries of the Province of Pennsylvania, which I omitted as not thinking it necessary in a paper the intention of which was only to reduce to a certainty by writing what in conversation I had delivered *viva voce*.[1]

But during this delay, the Assembly having prevailed with Governor Denny to pass an act taxing the proprietary estate in common with the estates of the people, which was the grand point in dispute, they omitted answering the message.

"*An Act Taxing the Proprietary Estate . . .*"

WHEN this act however came over, the Proprietaries, counseled by Paris, determined to oppose its receiving the royal assent. Accordingly, they petitioned the King in Council, and a hearing was appointed in which two lawyers were employed by them against the act, and two by me in support of it. They alleged that the act was intended to load the proprietary estate in order to spare those of the people, and that if it were suffered to continue in force and the Proprietaries who were in odium with the people, left to their mercy in proportioning the taxes, they would inevitably be ruined. We replied that the act had no such intention and would have no such effect. That the assessors were honest and discreet men under an oath to assess fairly and equitably, and that any advantage each of them might expect in lessening his own tax by augmenting that of the Proprietaries was too trifling

[1] By word of mouth.

to induce them to perjure themselves. This is the purport of what I remember as urged by both sides, except that we insisted strongly on the mischievous consequences that must attend a repeal, for that the money, £100,000, being printed and given to the King's use, expended in his service and now spread among the people, the repeal would strike it dead in their hands to the ruin of many, and the total discouragement of future grants, and the selfishness of the Proprietors in soliciting such a general catastrophe merely from a groundless fear of their estate being taxed too highly was insisted on in the strongest terms. On this Lord Mansfield, one of the counsel, rose, and, beckoning me, took me into the clerk's chamber while the lawyers were pleading, and asked me if I was really of opinion that no injury would be done the proprietary estate in the execution of the act. I said certainly. "Then," says he, "you can have little objection to enter into an engagement to assure that point." I answered, "None at all." He then called in Paris, and after some discourse his lordship's proposition was accepted on both sides; a paper to the purpose was drawn up by the Clerk of the Council, which I signed with Mr. Charles, who was also an Agent of the Province for their ordinary affairs, when Lord Mansfield returned to the Council Chamber, where finally the law was allowed to pass. Some charges were, however, recommended, and we also engaged they should be made by a subsequent law, but the Assembly did not think them necessary; for one year's tax having been levied by the act before the order of Council arrived, they appointed a committee to examine the proceedings of the assessors, and on this committee they put several particular friends of the Proprietaries. After a full inquiry they unanimously signed a report that they found the tax had been assessed with perfect equity.

The Assembly looked into my entering into the first part of the engagement as an essential service to the Province, since it secured the credit of the paper money then spread over all the country. They gave me their thanks in form when I returned. But the Proprietaries were enraged at Governor Denny for having passed the act, and turned him out with threats of suing him for breach of instructions which he had given bond to observe. He, however, having done it at the instance of the General and for His Majesty's service, and having some powerful interest at court, despised the threats and they were never put in execution.

[*Franklin's* Autobiography *ends here. An account of the remaining thirty years of his life will be found on the following pages.*]

The Continuation of Benjamin Franklin's Life
(1760-1790)

By Alan V. McGee, Ph.D.

When the Assembly succeeded in taxing the lands of the Penn family, they had established a precedent in their struggle to gain control over taxation within the Colony. They were so pleased with Franklin's success in this affair that they insisted on his remaining in London.

It was well for Britain that Franklin stayed. Helpful as he was to the Americans, he was as good a friend of Britain. Though Canada had surrendered in 1760, there was much doubt in England as to whether Britain should keep it as the spoils of war, since some of the French West Indies offered much more immediate profit to traders. Franklin's knowledge of the colonial trade and of the American wilderness enabled him to see its future value both to the Colonies already established and to the mother country. With greater wisdom than any other man in England, he had the largest vision of the future of empire. It was said that his writing on the subject induced the British to keep Canada when the treaty was signed.

He printed in the *London Chronicle* in 1760 his *Observations concerning the Increase of Mankind*, which he had written earlier in America, showing what promise of wealth there was for Britain in the increase of population in the Colonies. People had argued that emigration to America would impoverish the mother country. In stating the argument that, on the contrary, Britain would become richer, he wrote: "There is no bound to the prolific nature of plants or animals, but what is made by their crowding and interfering with each other's means of existence. Were the face of the earth vacant of other plants, it might be gradually sowed and overspread with one kind only, as, for instance, with fennel; and, were it empty of other inhabitants, it might in a few ages be replenished from one nation only, as, for instance, with Englishmen. Thus, there are supposed to be now upwards of one million English souls in North America (though it is thought scarce eighty thousand have been brought over sea) and yet perhaps there is not one the fewer in Britain, but rather many more, on account of the employment the Colonies afford to manufacturers at home. This million doubling, suppose but once in twenty-five years, will, in another century, be more than the people of England, and the greatest number of Englishmen will be on this side the water. What an accession of power to the British Empire by sea as well as by land!"

The justice of this prediction is seen from the fact that in 1775 there were three million people in the Colonies and that by 1800 there were 5,300,000; whereas in all Britain and Ireland there were but fifteen million. It was only in the permanent union of these peoples on the opposite shores of the Atlantic that Franklin foresaw future happiness and greatness. He wrote to Lord Kames that he had "long been of the opinion that the foundations of the future grandeur and stability of the British Empire lie in America; and though, like other foundations, they are low and little now, they are, nevertheless, broad and strong enough to support the greatest political structure that human wisdom ever yet erected. . . . All the country from the St. Lawrence to the Mississippi will in another century be filled with British people. Britain itself will become vastly more populous by the immense increase of its commerce; the Atlantic sea will be covered with your trading ships; and your naval power, thence continually increasing, will extend your influence round the whole globe, and awe the world."

Looking on the misery of the industrial population of Britain, Franklin foresaw no possibility of the development of an industrial civilization in America until the country had been thickly populated and the last frontiers had been settled. "England might quiet her fears," he said, "that the Colonies should take to manufacturing. Manufactures are founded in poverty. It is the multitude of poor without land in a country, and who must work for others at low wages or starve, that enables undertakers to carry on a manufacture and afford it cheap enough to prevent the importation of the same kind from abroad and to bear the expense of its own exportation. But no man who can have a piece of land of his own sufficient by his labor to subsist his family in plenty is poor enough to be a manufacturer and work for a master. Hence, while there is land enough in America for our people, there can never be manufactures to any amount or value."

He declared that it would be impossible for the Colonies to unite and rebel against the mother country, but "when I say such a union is impossible, I mean, without the most grievous tyranny and oppression. People who have property in a country, which they may lose, and privileges, which they may endanger, are generally supposed to be quiet, and even to bear much rather than hazard all. While the government is wise and just, while important civil and religious rights are secure, such subjects will be dutiful and obedient. *The waves do not rise but when the winds blow.* What such an administration as the Duke of Alva's in the Netherlands might produce, I know not; but this, I think, I have a right to deem impossible."

Franklin's thorough knowledge and his sharp, practical mind enabled him to formulate ideas which were to be as important to his cause as any political victories he might ever win in conflict with ministers. He was one of the British *literati* and a friend of Britain's leading thinkers. Adam Smith used to bring new chapters of *The Wealth of*

Nations to him for discussion before casting them into final form. His rank as a thinker was in itself an important contribution to the American cause.

His greatness had long been recognized in Europe, and it was chiefly with other men of learning that he associated in London. In 1759, St. Andrews had given him the degree of Doctor of Laws; Oxford followed in 1762 with the doctorate of Civil Law. These honors, primarily due to his scientific experiments, were nevertheless a sign of the high esteem in which he was held. Though the political and economic doctrines which he formulated during these years in London and Paris were important, the mere fact that he uttered them was of great value to America. He won much for her by his wise and good-humored negotiation. He won more by his greatness of mind.

In spite of the larger interests to which he now devoted himself, Franklin never gave up the detailed investigation of problems which came to his notice. He demonstrated that black clothing was warmer than light-colored clothing by showing that sun melted the snow faster under patches of black cloth. He perfected and performed on his harmonica, an instrument of musical glasses for which both Mozart and Beethoven composed. He wrote on the superior harmony of Scotch tunes composed for the harp. Whenever he saw any benefit to humanity from any discovery, he was quick to write on it, as he did in a persuasive introduction to a treatise on inoculation against smallpox, of which his second son had died. Though he still lacked time for systematic study, his inexhaustible energy kept him abreast of the work of his colleagues in the scientific societies of Europe; and when he met them on his trips to the continent, he matched his observations with theirs. His reputation as a scientist was world-wide.

In November, 1762, he was at last free to make the long-anticipated journey home. He loved his life in London, but said of it: "I feel here like a thing out of its place, and useless because it is out of place. How then can I any longer be happy in England. . . . I must go home." It was a real love that he was leaving. He wrote from Portsmouth before boarding his ship: "The attraction of reason is for the other side of the water, but that of inclination will be for this side. You know which usually prevails. I shall probably make this one vibration and settle here forever. Nothing will prevent it if I can, as I hope, prevail with Mrs. F. to accompany me, especially if we have a peace."

The first actions of his two-year stay in America were to go to New Jersey to see his son William received as Royal Governor of the Province, and to make a journey of over sixteen hundred miles in management of the post office, which took him to Virginia, to New York, and to his old friends and home in New England. He was also re-elected as usual to the Pennsylvania Assembly, now chiefly concerned with Indian raids against isolated settlements in the outermost parts of the Colony.

The raids had been part of the hostilities in the war against the French, and should have ceased when Canada submitted in 1760. The peace was not signed until 1763, however, some time after Franklin's slow voyage to America in a British convoy. When the official cessation of the war did not stop the raids, the Pennsylvanians were stirred to revenge on all Indians. An angry mob in Lancaster County known as the Paxton Boys turned riotously on the harmless Indians living in their midst and cruelly massacred them. Since the preservation of order required the prosecution of the rioters, and the commonwealth had at the same time to protect the Indians who had long lived peacefully nearer the Delaware, a situation of the utmost difficulty presented itself to the new Governor, young John Penn. In his dilemma, he was forced to rely upon Franklin, whose guidance he followed as the mob marched on Philadelphia to attack the Indians sheltered there under protection of the Assembly. "Governor Penn," Franklin wrote, "did everything by my advice, so that for about forty-eight hours I was a very great man, as I had been once some years before in a time of public danger." In spite of the fact that his good sense and bravery had protected the city, the populace sympathized with the mob and condemned the magnanimous conduct of Franklin and the Quakers in the Assembly. "Having turned [the mob] back and restored quiet to the city, I became a less man than ever; for I had by this transaction made myself many enemies among the populace."

Franklin was an admirer of the civilization of the Indian, and never showed himself more clearly a citizen of the world than in his eloquent *Narrative of the Late Massacres in Lancaster County*, in which he argued that revenge for the crimes of one tribe did not require revenge against all Indians. The people were too much frightened by the raids to be just, and the Governor was ill-disposed toward Franklin and the Quakers, so that neither was grateful to him for his preservation of the city when the mob had been about to attack. Governor Penn, a descendant of William Penn, offered large bounties for Indian scalps which, of course, bore no evidence of having been removed from hostile Indians. The scalp of a female brought fifty dollars. No more barbarous measures could have been taken to encourage the cruelty of the citizens or to enrage Franklin at the inhumanity of the later generations of the Penn family.

The Governor was also seeking to relieve the Penns of taxation, and, as if in retaliation for Franklin's and the Assembly's victory in 1760, was demanding that the proprietary lands should be taxed at the lowest rate prevailing in the Colony. Since the proceeds of the new tax bill were to be used to defray the cost of an expedition against the more distant Indians, who had really been responsible for the raids, the Governor's insistence was very irritating. It was at this time that Franklin wrote his satirical inscription for a monument to the sons of William Penn, which contrasts the wisdom and benevolence of the original Proprietor with the meanness and cruelty of his descendants. It concluded with the following bitter lines:

> 'To gain this point,
> They refused the necessary laws
> For the defense of their people
> And suffered their colony to welter in its blood,
> Rather than abate in the least
> Of these their dishonest pretensions.
> The privileges granted by their father,
> Wisely and benevolently
> To encourage the first settlers of the Province,
> They,
> Foolishly and cruelly,
> Taking advantage of the public distress,
> Have extorted from the posterity of those settlers;
> And are daily endeavoring to reduce them
> To the most abject slavery;
> Though to the virtue and industry of those people
> In improving their country,
> They owe all they possess and enjoy.
> A striking instance
> Of human depravity and ingratitude;
> And an irrefragable proof,
> That wisdom and goodness
> Do not descend with an inheritance;
> But that ineffable meanness
> May be connected with unbounded fortune.

Feeling rose so high that the Speaker resigned and Franklin was elected to replace him. Despairing of any future cooperation with the Penns or of any successful legislation, the Assembly petitioned the King to remove the Proprietors and take the Colony directly under royal control. The Penns, allying themselves with the conservative voters in Philadelphia, sought revenge against Franklin and in a closely contested election on the issue caused him to lose his seat. Their revenge was not entirely successful, however, since Franklin's party still dominated the Assembly, and he was chosen as agent to take the petition to England for submission to the Crown. It was not inclination, therefore, but reason and necessity which within two years sent Franklin back to the cultivated society of London. His still high opinion of the new King, George III, led him to think that royal protection would redress the grievances of his fellow-citizens. In this sublime but foolish hope, he sailed from Philadelphia on November 7, 1764.

When he reached London on December 10, however, a far more important problem confronted him. The Stamp Act had been proposed. Far more dangerous to the future union of the British peoples than the quarrel with the Penns was this new effort to tax with revenue stamps all important transactions within the Colonies. No more oppressive or indeed unenforceable tax could have been levied, and therefore Franklin devoted himself to the defeat of the bill. It was not until a year after his arrival in England that he was able to present his petition to free Pennsylvania of its Proprietors. It was not granted; not until after the Declaration of Independence was signed did the Legislature buy out the Penns with a grant of £130,000.

Though Franklin failed to prevent the passage of the Stamp Act, he was willing to acquiesce in the imposition of the tax with the hope that other repressive measures would be lifted and that in time, when its folly was seen, even this law would be repealed.

Neither he nor anyone else in England expected the violent storm of protest to the Act which arose in America. Passion ran so high that Franklin was accused of having formulated it and of having profited personally by its adoption. Forced to defend himself against this criticism, he pointed out that the tax fell especially heavily on him as the publisher of a newspaper, every copy of which was taxed one halfpenny if printed on a half-sheet of paper and one penny if printed on a full sheet. "Depend upon it," he wrote, "I took every step in my power to prevent the passage of the Stamp Act. No one could be more concerned in interest than myself to oppose it sincerely and heartily. But the tide was too strong against us. . . . We might as well have hindered the sun's setting. That we could not do. But since 'tis down, my friend, and it may be long before it rises again, let us make as good a night of it as we can. We may still light candles. Frugality and interest will go a great way toward indemnifying us. Idleness and pride tax with a heavier hand than kings and parliaments; if we can get rid of the former, we may easily bear the latter."

Though he was alarmed by the violence of the colonial response, and though he had said that the Americans should "keep within the bounds of moderation and obedience" as "the only way to lighten and get clear of [their] burdens," the Act had nevertheless brought about the first signs of that colonial unity which he had advocated in 1754 at Albany. For once, however, the Americans were now more radical than he, though his own proposals—the representation of the Colonies in Parliament, or a parliament for the Colonies under the Crown—so conservative in themselves, were still too radical for the English.

While it was the American boycott of English manufactures which brought about the repeal of the Stamp Act in 1766, it was Franklin who provided the climactic arguments for its defeat. He appeared at the bar of the House of Commons to be cross-questioned by enemies who were but clumsy defenders of a lost cause and by friends who were master-propagandists. With masterly tact and agile wit he fended off all attacks, avoided the excessive levity of his more enthusiastic friends, and taught the Commons more than they had ever known about the Colonies. He produced the final evidence of America's self-sufficiency and demonstrated its inflexible determination not to yield to the tax. He said that he would rather leave uncollected the debts owed to him than collect them under the terms of the Act, and he concluded his testimony with a well-rehearsed question and answer:

"What used to be the pride of the Americans?"

"To indulge in the fashions and manufactures of Great Britain."

"What is now their pride?"

"To wear their old clothes over again until they can make new ones."

So he redeemed himself from all the criticism to which he had been subject in America, and thenceforth appeared to his fellow countrymen only as a hero. Even his opponents in Pennsylvania were glad that they had defeated him in the last election to the Assembly so that he had been free to go to London.

The Americans had made a distinction between internal taxes, those levied on property or transactions within the Colonies, and external taxes, the duties imposed on the trade between the Colonies and the mother country. The distinction was not wholly logical, for it opposed one kind of taxation without representation and allowed another. Franklin said in his examination in the House: "The sea is yours; you maintain, by your fleets, the safety of navigation on it and keep it clear of pirates: you may have therefore a natural and equitable right to some toll or duty on merchandise carried through that part of your dominions, toward defraying the expense you are at in ships to maintain the safety of that carriage." In case the duty is felt to be unjust or excessive and if "people do not like [an article] at that price, they refuse it; they are not obliged to pay it. But an internal tax is forced from the people without their consent, if not laid by their own representatives." And in Franklin's opinion, the Colonies imported no object, even cloth, that they could not do without.

One of his questioners suggested that there was no difference between the internal and external taxes and that the Americans should not accept the one and reject the other. He answered: "Many arguments have been used to show that there is no difference and that, if you have no right to tax them internally, you have none to tax them externally, or make any other law to bind them. At present they do not reason so; but in time they may be convinced by those arguments."

Though the Stamp Act was repealed only a week or two after Franklin's appearance in the House, Townshend, Chancellor of the Exchequer, set about the framing of new acts which should lessen the authority of the colonial legislatures and reassert Parliament's right to tax. Part of the new levy was to be used to pay the salaries of the colonial governors, thus freeing them from dependence on their individual assemblies. As if to emphasize the true purpose of the Acts, the New York Assembly was suspended at the same time for having refused to vote money to support the British army in America. The taxes themselves, though collected at the ports, were plainly for revenue of the same kind as the Stamp Act had been—to support the British soldiers in the Colonies. They were to be collected in the colonial ports in order to placate those Englishmen who wanted Parliament to go as far as possible in the direction of an internal tax.

The Townshend Acts were foolish and blundering; for their purpose was apparent. It was equally apparent that the Colonies would not submit to them. Events moved on unhappily to the Virginia Resolves and finally, on March 5, 1770, the very day of the repeal of the Acts, to the Boston Massacre, in which a squad of British soldiers fired into a crowd and killed four men.

Franklin had foreseen these dangers, as he wrote before the passage of the Acts: "In the meantime, every act of oppression will sour their tempers, lessen greatly—if not annihilate—the profits of your commerce with them, and hasten their final revolt; for the seeds of liberty are universally found there, and nothing can eradicate them." It was evident that the colonists now understood that the economic affairs of America could not depend on a government across the Atlantic which could impose monopolistic terms of trade or could set oppressive terms of bargaining. Liberty for America required economic as well as political freedom from Europe.

Writing as an anonymous Briton, Franklin said in the *London Chronicle*: "Iron is to be found everywhere in America, and the beaver furs are the natural products of that country. Hats, and nails, and steel are wanted there as well as here. It is of no importance to the common welfare of the empire whether a subject of the King obtains his living by making hats on this or on that side of the water. Yet the hatters of England have prevailed to obtain an act in their own favor restraining that manufacture in America in order to oblige the Americans to send their beaver to England to be manufactured, and purchase back the hats loaded with the charges of a double transportation. In the same manner have a few nail-makers and a still smaller body of steel-makers (perhaps there are not half a dozen of these in England) prevailed totally to forbid by act of Parliament the erection of slitting-mills, or steel-furnaces in America. . . . The whole American people were forbidden the advantage of the direct importation of wine, oil, and fruit, from Portugal, but must take them loaded with all the expense of a voyage one thousand leagues round about, being to be landed first in England, to be reshipped for America; expenses amounting, in war time at least, to thirty pounds per cent. more than otherwise they would have been charged with; and all this, merely that a few Portugal merchants in London may gain a commission of those goods passing through their hands."

During all these years, indeed, Franklin's greatest function was the explanation of America both to herself and to the Old World. His value became evident to the Colonies, and though there was some ill-will against him because of his moderation, he was made Agent for Georgia as well as Pennsylvania in 1768, for New Jersey in 1769 and Massachusetts in 1770. These new posts were the more agreeable because his partnership with Hall in the printing business had been dissolved in 1766, and he no longer had income from that source. He had moved from business to the world of diplomacy and was now an unofficial ambassador from all the Colonies. The ministers from other countries, especially France, began to take an interest in him as a potential ally in any future attempts to weaken the British power.

But he was to suffer many humiliating defeats before he was to return to America with all hope of union gone. In January, 1770, when he presented his credentials as Agent of the Massachusetts Assembly, Lord Hillsborough, Secretary of State for Colonies, rejected them, saying that he could accept an appointment only through an act signed by the Governor of the Colony. Since Franklin was an Agent of the Assembly and not of the Governor, who was himself an Agent of the Crown, this was a direct refusal to deal with the people of Massachusetts. Franklin then said to Hillsborough, "It is, I believe, of no importance whether the appointment is acknowledged or not, for I have not the least conception that an Agent can *at present* be of any use to any of the Colonies." Hillsborough later remarked that Franklin seemed to believe that the Colonies could expect neither favor nor justice during his administration. "I find he did not mistake me," Franklin said.

The greatest humiliation he suffered was the abusive address of Wedderburn, counsel for Governor Hutchinson of Massachusetts, at the meeting of the House of Lords' Committee on December 11, 1773, to examine the petition of the Massachusetts Assembly for the Governor's removal. Franklin had sent to friends in Massachusetts certain letters written to Ministers of the Crown by Hutchin son, urging stricter rule of the Colonies. He refused to reveal how he got them, and Wedderburn did not hesitate to call him a thief. Throughout the whole shameful harangue, Franklin stood quiet, dressed in a figured blue velvet suit, disdaining to answer or be questioned. The abuse was also prologue to his dismissal as postmaster.

In the face of all this, Franklin made no comment. As he wrote to a friend, "Such censures I have generally passed over in silence, conceiving when they were just, that I ought rather to amend than defend, and when they were undeserved, that a little time would justify me. Splashes of dirt thrown upon my character I suffered while fresh to remain. I did not choose to spread by endeavoring to remove them, but relied on the vulgar adage that they would all rub off when they were dry." Of his loss of the post office, he said merely, "I have lost a little place that was in their power, but I can do very well without it."

No one could then have expected it, but before long he was to head the post office of the United States, and still later, in 1778, he was to wear his figured blue velvet suit in Paris at the signing of the first treaty between France and the independent states.

The weakening of the ties between Britain and the Colonies was hastened by the progressive blunders of the King and his Ministers. The monopoly on tea given to the East India Company threatened to ruin many American merchants and incited the men of Boston to the famous Tea Party.

Franklin's friendship and collaboration with the Opposition could achieve nothing in the then corrupt state of Parliament. He accomplished more by the use of his pen. His *Rules by Which a Great Empire May Be Reduced to a Small One* merely listed the steps which had been taken by the administration to drive the Colonies to rebellion. His most amusing satire was the *Edict by the King of Prussia*. Recalling that England had been settled from Germany, he applied to the Prussian King the motives and actions of George III toward America:

"Whereas it is well known to all the world that the first German settlements made in the Island of Britain were by colonies of people, subject to our renowned ducal ancestors and drawn from their dominions under the conduct of Hengist, Horsa, Hella, Uff, Cerdicus, Ida, and others; and that the said colonies have flourished under the protection of our august house for ages past, have never been emancipated therefrom, and yet have hitherto yielded little profit to the same.

"And whereas we ourself have in the last war fought for and defended the said colonies against the power of France and thereby enabled them to make conquests from the said power in America, for which we have not yet received adequate compensation.

"And whereas it is just and expedient that a revenue should be raised from the said colonies in Britain towards our indemnification, and that those who are descendants of our ancient subjects and thence still owe us due obedience should contribute to the replenishing of our royal coffers as they must have done had their ancestors remained in the territories now to us appertaining:

"We do therefore hereby ordain and command that, from and after the date of these presents, there shall be levied and paid to our officers of the customs on all goods, wares, and merchandises, and on all grain and other produce of the earth exported from the said Island of Britain, and on all goods of whatever kind imported into the same, a duty of four and a half per cent *ad valorem* for the use of us and our successors. And that the said duty may more effectually be collected, we do hereby ordain that all ships or vessels bound from Great Britain to any other part of the world or from any other part of the world to Great Britain shall in their respective voyages touch at our port of Koningsberg, there to be unladen, searched, and charged with the said duties."

The edict goes on to list other British laws for the Colonies and even to cite the precedent of the British Parliament for the validity of its own attitude.

Since Franklin's importance in America was now evident to all, the British resorted finally to efforts at secret negotiation with him, accompanied by offers of place and honor which he would be bound to find dishonorable. The death of Deborah Franklin in December, 1774, put an end to these futile secret interviews and made him resolve to go home. Until his departure he hoped that some honorable compromise might be achieved; but when he reached

America in 1775 he was convinced that nothing less than independence would serve.

During those difficult years in England when the fate of America rested on his shoulders, he was forced to devote most of his energies to politics. Yet his vitality had allowed him to continue his investigation into all sorts of scientific problems. He bought scientific instruments for Harvard College and entered into a scheme for sending European animals and seeds to the Pacific islands. He also had time for experiments with a phonetic alphabet, the ventilation of public rooms, the effect of the depth of canals upon the speed of boats, and the effect of oil in calming waves. He wrote on the common cold and on lead poisoning, the symptoms of which he was the first to trace to their actual cause. He was made a member of the Royal Society in 1756 and became a fast friend of Joseph Priestley and other men of science in England and on the continent. His scientific success was complete, except that King George III was no longer willing to believe in Franklin's theories about pointed lightning rods and persisted in authorizing the use of blunt ones only!

On May 5, 1775, just seventeen days after the battles of Lexington and Concord, Franklin arrived in Philadelphia, where he was hailed as the savior of America. On the following day he was elected a delegate to the Second Continental Congress. Coming as he did from the very center of the dispute, Franklin knew what extreme measures were necessary; in consequence, his political experience at home was not altogether happy. The Pennsylvania Assembly, of which he was also a member, was still conservative, and Franklin felt compelled to absent himself from the meeting at which it took the oath of fealty to the King. He was the oldest man and the wisest statesman in the Continental Congress, but he could not impress his will upon either the Pennsylvania delegation or the Assembly.

His personal life must have been equally difficult. His only son William Franklin, Governor of New Jersey, could not be persuaded even by his father to join the revolutionary cause. Though Franklin had brought William's son Temple from London and given him the first opportunity to know his father, he could not bear to allow the boy to remain under such Tory influence and therefore kept him in Philadelphia as private secretary.

In spite of his age Franklin lent all his energies to the new cause. He established the new post office and went to Cambridge to help Washington organize the new army. His most exhausting service was a futile mission to the Canadians, which exposed him to grave physical danger in the cold weather of the winter and early spring of 1776. Tired and unwell, he arrived in Philadelphia in June to take part in the deliberations leading to the Declaration of Independence, which he helped to revise and which he also signed.

Valuable as he must have been in legislative debate (he spoke seldom and briefly, but always to the point), he was still more valuable as a diplomat. Congress sent him among others to negotiate with Lord Howe, whom he had known in London, and on September 26, 1776, appointed him commissioner to deal with France. He put his son-in-law, Richard Bache, in charge of the post office, entrusted his papers, including the *Autobiography*, to an old Tory friend, and lent to the Congress all the money he could raise. On October 27 he sailed, taking with him two grandsons, Temple Franklin and Benjamin Franklin Bache. On that hazardous voyage, when capture would have meant his death, he devoted himself to observing the course and temperature of the Gulf Stream. Off the French coast his ship captured two British prizes.

Though he had come to France in the hope of enlisting the French government as an ally, he was unaware that all France was his personal ally. Excitement at the prospect of his arrival was only increased by his necessary rest on the way after the rough sea voyage. When he reached Paris, clad in a plain brown suit with white stockings and a beaver hat, all French society was delighted with the unassuming simplicity of the New World philosopher, whom they took for a Quaker. Fashionable ladies even dressed their hair in imitation of his beaver hats!

The French ministry could not recognize him openly since that would have been hostile to England. So under the cloak of educating his two grandsons he retired to Passy, then a suburb of Paris, where he carried on his official business. Though he could not appear officially in the diplomatic corps, he became a friend of its members and continued his friendship with French men of science.

Franklin's scientific interests were as broad as they had ever been, though he had less time to devote to investigation of his own. He was much interested in the theories of the Physiocrats, a group of French economists who maintained that only agriculture produced real wealth. Because he had found that manufacturing was successful only in the widespread poverty of Britain and Ireland, he agreed with the Physiocrats that an agricultural America would be a land of boundless prosperity.

After the war he was to write *Information to Those Who Would Remove to America*, both to encourage emigration and to warn the unfit against misconceptions of the new life they would lead there. In the meantime, his knowledge and his theories won support for the Colonies. His success in society and in the world of learning was the best recommendation an American ambassador could have had.

He must often have been far from happy himself. The long series of British successes in the war discouraged the French and made an open alliance impossible. Philadelphia was in the hands of the enemy, his house was occupied by British officers (Major André stole Benjamin Wilson's portrait of Franklin), his family were in flight, and his son William had been imprisoned in Connecticut. Not until some months after the surrender of Burgoyne at Saratoga on October 17, 1777, did the French make an alliance with the United States and recognize Franklin as Minister Plenipotentiary.

He was greatly overworked, his personal problems were innumerable, and his associates were often untrustworthy. He was incessantly bothered with all sorts of people, interested in his cause or desirous of going to America. Finally, in exasperation, he wrote for his own pleasure a form letter of introduction:

"The bearer of this, who is going to America, presses me to give him a letter of recommendation, though I know nothing of him, not even his name. This may seem extraordinary, but I assure you it is not uncommon here. Sometimes, indeed, one unknown person brings another equally unknown to recommend him; and sometimes they recommend one another! As to this gentleman, I must refer you to himself for his character and merits, with which he is certainly better acquainted than I can possibly be. I recommend him, however, to those civilities which every stranger of whom one knows no harm has a right to; and I request you will do him all the good offices and show him all the favor that on further acquaintance you shall find him to deserve. I have the Honor to be, etc."

It would be rash to believe that this letter was actually used. To relieve his feelings Franklin wrote many letters which he never sent, like the famous letter to Strahan, the English printer, with whom he nevertheless remained on friendly terms:

"Philadelphia, July 5, 1775.

"Mr. Strahan, You are a Member of Parliament, and one of that majority which has doomed my country to destruction.—You have begun to burn our towns, and murder our people.—Look upon your hands! They are stained with the blood of your relations!—You and I were long friends:—You are now my enemy,—and I am

"Yours,

"B. Franklin"

But Franklin's greatest trouble was with his jealous and suspicious associates. That he had deep feelings on the subject can be seen in a letter of encouragement he wrote to George Washington:

"Should peace arrive after another campaign or two, and afford us a little leisure, I should be happy to see your Excellency in Europe and to accompany you, if my age and strength would permit, in visiting some of its ancient and most famous kingdoms. You would on this side of the sea enjoy the great reputation you have acquired, pure and free from those little shades that the jealousy and envy of a man's countrymen and contemporaries are ever endeavoring to cast over living merit. Here you would know and enjoy what posterity will say of Washington. For 1000 leagues have nearly the same effect with 1000 years. The feeble voice of those groveling passions cannot extend so far either in time or distance. At present I enjoy that pleasure for you, as I frequently hear the old generals of this martial country (who study the maps of America, and mark upon them all your operations) speak with sincere approbation and great applause of your conduct, and join in giving you the character of one of the greatest captains of the age.

"I must soon quit the scene, but you may live to see our country flourish, as it will amazingly and rapidly after the war is over. Like a field of young Indian corn, which long fair weather and sunshine had enfeebled and discolored, and which in that weak state by a thunder gust of violent wind, hail, and rain seemed to be threatened with absolute destruction, yet the storm being past, it recovers fresh verdure, shoots up with double vigor, and delights the eye, not of its owner only, but of every observing traveler."

Though Franklin was not a Quaker, he hated war, and the happiest part of his stay in France came after the cessation of hostilities in 1781. He wrote: "Justice is as strictly due between neighbor nations as between neighbor citizens. A highwayman is as much a robber when he plunders in a gang as when single; and a nation that makes unjust war is only a great gang." Peace came officially when the treaty with the British was signed in September, 1783. "We are now friends with England and with all mankind," Franklin wrote Josiah Quincy. "May we never see another war! For in my opinion there never was a good war or a bad peace."

Now the British Ambassador called regularly and for

pleasure at Passy. Here Franklin held his little court of learned friends and beautiful and witty French ladies. The simple Pennsylvanian, an accomplished admirer of the graces, was living in the most polished society of Europe. His friendships with Mme. Brillon and Mme. Helvétius and his long mock-courtship of them have left correspondence of never-ending delight.

In 1785, two years after the treaty of peace, Congress allowed him to relinquish his post, in which Jefferson succeeded him. To the grief of all his French friends, whom he now knew better than those in America, he set out for home. A royal litter was provided to take him to the coast, and crowds of people paid homage along the way. His biographer, Carl Van Doren, contrasts his return to Philadelphia in September 1785 with that first voyage of the obscure young printer in 1726. "Now he came back the most famous private citizen in the world." The city's official reception lasted a week. Honors and duties were thrust upon him. "I had not firmness enough to resist the unanimous desire of my countryfolks," he wrote, "and I find myself harnessed again in their service. . . . They engrossed the prime of my life. They have eaten my flesh, and seem resolved now to pick my bones."

Three of his remaining five years he spent as President of Pennsylvania, though his ill-health and advancing years often caused him to be absent from his post. At the same time, he was active as a delegate in the committees of the Constitutional Convention, to which he contributed the essential compromise of the balance of power in the two houses of Congress.

It was not until 1788, at the age of eighty-two, that he retired from public life. In the quiet pleasure of his two last years he resumed work on the *Autobiography*, and kept up as well as he could his wide range of interests and his correspondence abroad. His heart turned from the simple Quaker city to the friends he had known in Paris before the French Revolution broke out. All the pleasures of retirement, he wrote, "do not make me forget Paris and the nine years' happiness I enjoyed there. . . . And now, even in my sleep, I find that the scenes of all my pleasant dreams are laid in that city or in its neighborhood."

During the last years of his life he suffered greatly from stone in the bladder, but remained cheerful. "For my own personal ease," he wrote President Washington, September 16, 1789, "I should have died two years ago; but, though these years have been spent in excruciating pain, I am pleased that I have lived them, since they have brought me to see our present situation"—the establishment of the new government with Washington as President. His mind was clear and active until death came quietly on April 17, 1790.

All his life Franklin had thought of himself as a printer. His will began: "I, Benjamin Franklin, Printer, late Minister Plenipotentiary from the United States of America to the Court of France, now President of Pennsylvania. . . ."

Perhaps the celebrated epitaph that he composed in 1728 was intended more seriously than is usually thought:

> The Body of
> B Franklin,
> Printer;
> Like the Cover of an old Book,
> Its Contents torn out,
> And stript of its Lettering and Gilding,
> Lies here, Food for Worms.
> But the Work shall not be wholly lost:
> For it will, as he believ'd, appear once more,
> In a new & more perfect Edition,
> Corrected and amended
> By the Author.

But the stone upon his grave in the burying ground of Christ Church, Philadelphia, where he lies beside his wife, bears the simple inscription:

> Benjamin and Deborah Franklin
> 1790

The Body of
B Franklin,
Printer;
Like the Cover of an old Book,
Its Contents torn out,
And stript of its Lettering and Gilding,
Lies here, Food for Worms.
But the Work shall not be wholly lost:
For it will, as he believ'd, appear once more,
In a new & more perfect Edition,
Corrected and amended
By the Author.
He was born Jan. 6 1706
Died   17

Epitaph written by Franklin when he was 22 years old. From the original in his own hand in the Yale University Library.